Waterborne Coatings and Additives

Waterborne Coatings and Additives

Edited by

D. R. Karsa and W. D. Davies

Akcros Chemicals UK Ltd, Manchester, UK

THE ROYAL
SOCIETY OF
CHEMISTRY

The Proceedings of a Symposium organized by The Royal Society of Chemistry and the Society of Chemical Industry, held on 14–15 September 1994, at Manchester Conference Centre, UMIST, UK

The cover diagram shows the 'house of cards' structure formed from hectorite clay platelets during gel formation. Clays are used as thickening agents in some coatings formulations. See Figure 6 on page 223.

Special Publication No. 165

ISBN 0-85404-740-9

A catalogue record for this book is available from the British Library

Published by The Royal Society of Chemistry,
Thomas Graham House, Science Park, Milton Road,
Cambridge CB4 4WF, UK

Printed in Great Britain by Hartnolls Ltd., Bodmin

Preface

The paint and surface coatings industry continues to be influenced by sustained environmental pressures, in particular the reduction or elimination of volatile organic compounds (VOCs). This in turn has further accelerated the growth of waterborne formulations. Not only has the move been away from solvent to water-based coatings, but there has also been environmental pressure to eliminate coalescing solvents and glycol-based freeze-thaw stabilisers from established emulsion paints. In some instances this has led formulators to consider total reformulation of their copolymers in order to retain or optimise the film forming properties of their paint in the absence of such additives.

At the same time renewed interest has been observed in the development of water-based emulsions of high performance, film forming resins such as alkyds, epoxies, polyesters and polyurethanes.

In 1988, the North West Industrial Division of the Royal Society of Chemistry organised a two day symposium at Liverpool University on "Additives for Water-based Coatings" (RSC Special Publication No. 76, ISBN 0-85186-607-7). This present symposium, 6 years later, attempts to outline the further developments that have occurred in waterborne polymers and additives since then.

The development of anti-pollution legislation is presented and a range of water-based resins and their end-use applications are considered in depth. These include metal treatment, concrete and textile applications. Cross-linking mechanisms are also reviewed and specific papers consider amino resins, waterborne radcure coatings and water-based urethanes. The latter half of the proceedings considers a range of additives currently used in the manufacture and formulation of aqueous systems. These include biocides and foam control agents, acetylenic and polymeric surfactants, clay rheology modifiers, driers and metallic pigments.

The subject of water-based coatings and the many additives used in their formulation is too broad to cover in a single volume. However, the editors hope that this monograph will prove to be a valuable addition to current literature on this topic, suitable for both readers relatively new to the field as well as experienced workers in the coatings industry.

D.R. Karsa
March 1995

Contents

Additives

Applications and Performance

Waterborne Polymers: Design for Performance

J. B. Clarke and E. Alston

ALLIED COLLOIDS LIMITED, PO BOX 38, CLECKHEATON ROAD, LOW MOOR, BRADFORD, WEST YORKSHIRE BD12 0JZ, UK

1. INTRODUCTION

Waterborne coatings were first introduced during the 1930's when commercial paints based on a polyvinyl acetate latex were developed in Canada.

As synthetic resin technology improved and the commercial availability of suitable monomers increased, latices based on acrylics, styrene-butadiene, styrene-acrylics, vinyl acetate - acrylics, vinyl acetate - versatate and vinyl acetate-ethylene have been specifically designed for water based coating formulations. Such formulations find wide application in a number of industries eg textile, paper, adhesives as well as paints.

Water based paints initially found their niche in the DIY field where the concepts of no solvent odours and the convenience of cleaning brushes and rollers in water were successfully marketed. The further development to replace solvent systems for the more demanding industrial applications has been somewhat retarded due to technical deficiencies in the raw materials.

However, in recent years the ever increasing legislation and attitudes towards reducing and eliminating the release of volatile chemicals into the atmosphere has put the coatings formulator under increased pressure. Whilst a number of alternative products to solvent based are available, eg powder coatings, radiation curable systems, waterborne paints retain the conventional characteristic of being liquid and can consequently be applied using existing equipment.

Fig 1 summarises UK paint sales in the period 1970-1990.

In 1970 46% of total was water based, which increased to 60% in 1990.

Fig 1. UK Paint Sales

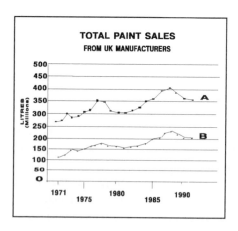

A = total paint sales   B = water based

2.   **THE DESIGN OF FORMULATIONS FOR HIGH PERFORMANCE**

A number of ideal requirements may be listed. (Table 1)

The quest for optimum performance relies to a large extent on the design of the base film former. However, the mixture of resin and pigment alone does not produce a coating.   Other essential ingredients must be incorporated to achieve design performance.   (Table 2)

Table 1

Ideal Requirements for High Performance

- Elimination of all hazardous volatiles
- Elimination of unpleasant odours - not necessarily hazardous
- Ambient temperature film forming and curing
- Maximisation of resistance properties
  - water
  - alkali
  - solvents
  - heat
  - blocking
- Stable to storage conditions
- Suit the application method
- Cost effective

Table 2

Essential Additives

- Pigment dispersing agent
- Thickener/rheology modifier
- Biocide
- Foam control agents
- Slip and mar additives
- Adhesion promoters

The selection and optimisation of additives is of equal importance to the selection of the main binder.

In order to exemplify designing for performance the following case study has been selected.

3.   **The design and formulation of an ambient cure wood lacquer**

3.1. MARKET SURVEY

Before any research project is commenced a thorough market
survey of the potential for the proposal new product is
essential. An ambient curing wood lacquer would find application
in both industrial and retail outlets.

Table 3

Industrial wood finishes in Europe 1982-1992

	1982		1987		1992	
	TONS	%	TONS	%	TONS	%
Nitrocellulose	152,200	39.4	134,650	31.0	107,280	25.0
Acid-cured	132,115	34.2	150,245	34.6	122,235	28.4
Polyurethanes	30,900	8.0	66,445	15.3	91,165	21.2
Polyesters	28,975	7.5	34,750	8.0	42,350	9.9
Rad-cured	11,590	3.0	23,000	5.3	38,680	3.2
Waterborne	3,865	1.0	7,385	1.7	13,725	3.3
All others	26,665	6.9	17,800	4.1	14.115	3.3
TOTAL	386,310	100.	434,275	100.	429,550	100.

It can be seen that the demand for the more traditional
resin systems is in decline to the benefit of those more
environmentally acceptable. The rate of growth of waterborne
systems is particularly impressive. The 1992 figures are broken
down for the major European manufacturers in table 4 from which
it can be seen that Germany is currently the major user of waterborne industrial wood finishes.

Table 4. National European Demands For Industrial Wood Coatings By Type Of Resin System (1992)

Major Markets	Nitrocellulose	Polyurethanes	Acid-cured	Polyesters	Rad-cured	Waterborne	All others	Total
France	11,830	4,440	3,450	1,430	860	865	1,775	24,650
Germany	37,825	18,690	1,006	4,450	6,140	8,365	3,470	89,000
Italy	11,800	64,125	2,565	25,650	23,650	900	135	128,250
Spain	11,430	20,300	5,485	3,660	1,830	455	2,565	45,725
United Kingdom	8,215	1,150	9,665	475	955	890	375	21,725

An assessment of the retail market for woodcare products for the whole of Europe is difficult due to the different reporting systems in the various countries but detailed figures for the U.K. market are available:-

Table 5. UK Retail Woodcare Products - Volume

UK RETAIL VOLUME MARKET FOR WOODCARE PRODUCTS BY PRODUCT SECTOR 1989 - 1993 (MILLIONS OF LITRES)					
PRODUCT SECTOR	1989	1990	1991	1992	1993
PRESERVATIVES	17.2	18.0	20.3	19.4	21.6
VARNISHES, INTERIOR	3.6	3.5	3.7	3.9	3.7
VARNISHES, YACHT	1.4	1.5	1.4	1.3	1.3
STAINS	3.0	2.8	3.2	3.3	3.2
DYES	0.8	1.2	1.4	1.1	1.2
TOTAL	26.0	27.0	30.0	29.0	31.0

Table 6. UK Retail Woodcare Products - Value

UK RETAIL VALUE MARKET FOR WOODCARE PRODUCTS BY PRODUCT SECTOR 1989 - 1993 (£ MILLION)					
PRODUCT SECTOR	1989	1990	1991	1992	1993
PRESERVATIVES	22.6	24.0	25.0	26.8	24.8
VARNISHES, INTERIOR	25.4	26.0	29.5	34.7	33.8
VARNISHES, YACHT	3.9	4.0	4.8	5.9	5.7
STAINS	18.5	21.0	23.8	26.3	29.8
DYES	2.6	5.0	2.9	6.3	8.9
TOTAL	73.0	80.0	86.0	100.0	103.0

These two tables are represented graphically in Fig 2.

Fig 2. UK Retail market for wood coating products 1993
(% of total volume and value)

Volume

Value

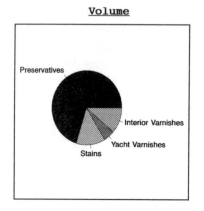
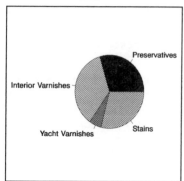

Source: Trade and MSI estimates

Estimates show that in 1993 waterbased products counted for 55% of the market in fencing treatment, 27% of the interior varnish market and 18% of all woodstains. The latter two segments are increasing at 20% per annum currently and account for the majority of sales growth.

Users of D.I.Y. products are taking environmental issues increasingly seriously although product effectiveness and performance will still outweigh environmental concerns in the final analysis.

Leading U.K. manufacturers have introduced "solvent free" and "low odour" products and their range of waterbased alternatives to solvent based coatings has vastly increased. Although the recent U.K. recession has reduced D.I.Y activity, market sources are confident that growth in the waterbased sector will continue to improve as the overall market recovers.

Having established a need for improvements in the performance of waterbased coatings in a growth sector of the market the next step was the design an appropriate polymer system.

3.2 Factors affecting the Design of an Ambient Cure Resin

The most widely used ambient temperature curable solvent based resin systems are based on convertible resins which are usually fully soluble in solvent. On drying, at ambient temperature, the simple elimination of solvent by evaporation is accompanied by a chemical conversion reaction which develops the required resistance and durability properties.

Conventional air drying latices, on the other hand, do not undergo chemical conversion during the drying process.

The first mechanistic model of the drying process of latices was published by Vanderhoff et al in 1973[1]. A three stage model was proposed (Fig 3).

Fig 3. Film Formation From Emulsions

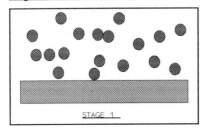

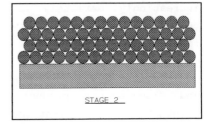

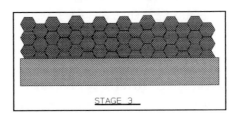

In the first stage the water evaporates from the surface, gradually concentrating the system to a point where the particles begin to contact one another which leads to the second stage.

Particle coalescence then begins to take place, leading to stage 3 when the film is considered dry and fully coalesced.

The mechanism has been studied further by other workers and reported in the literature. [2,3]

The performance of conventional latices is largely governed by three factors. (Table 7)

Table 7
Factors Governing Latex Performance

Polymer Composition
Molecular Weight
Particle Size

These three factors in their turn govern the glass transition temperature (Tg) and the minimum film forming temperature (MFFT) which are the essential parameters requiring optimisation for maximum performance.

Table 8 lists the Tg values of a range of monomers used by latex manufacturers to generate water based coating binders.

Table 8
Tg Values of a Selection of Common Monomers

POLYMER	Tg°C
METHYL ACRYLATE	6
ETHYL ACRYLATE	-24
N-BUTYL ACRYLATE	-55
2-ETHYL HEXYL ACRYLATE	-60
METHYL METHACRYLATE	105
BUTYL METHACRYLATE	20
STYRENE	100
ACRYLONITRILE	105
VINYL ACETATE	28

In order to achieve maximum resistance properties a high glass transition temperature is normally required. Unfortunately high Tg is normally accompanied by high MFFT. This conventionally is overcome by using volatile coalescing agents to promote the initial film formation.

To achieve ambient temperature film formation, polymers can be engineered using the 'core-shell' method developed some years ago.

Fig 4. Core-shell Emulsion Particle

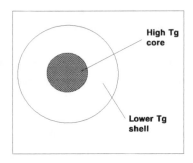

Whilst this technique produces the lower MFFT requirement to allow reduction or even elimination of volatile coalescents it does not provide the mechanism for achieving high performance.

In order, therefore, to bridge this shortfall methods of raising the Tg during the drying stage have been very actively researched in recent times.

3.2.1 AMBIENT TEMPERATURE CROSSLINKING SYSTEMS

Ambient temperature crosslinking systems contain internal moieties, that cause a crosslinking reaction to occur upon drying, after the product has been applied to the substrate. This type of system requires no post addition of catalyst immediately prior to application and no thermal curing program.

Certain multifunctional additives are introduced into the monomer recipe to functionalise the resultant polymer chains allowing crosslinking between functionalities to occur as the coating dries.

A judicious selection of polymer modifiers is required, so that a crosslinking system can be manufactured which will undergo crosslinking reactions at room temperature. A further criteria is set in that the crosslinking reactions should not take place instantly but at the required time i.e. upon drying. The polymerisable crosslinking moieties must be able to withstand the polymerisation conditions without reacting. Otherwise the film forming properties of the end product may be degraded.

The optimum crosslinking system should react by loss of water via a condensation reaction as the polymer dries. Therefore the faster the polymer dries, the more quickly the crosslinking will take place. It can be seen that such systems are ideally suited for use in coatings, as drying of thin films is normally rapid.

The use of this type of system gives products of similar application properties (viscosity, solids, rheology) to current water based formulations, whilst giving a final film of excellent all round performance.

As the crosslinking progresses, the glass transition temperature (Tg) of the polymer increases, giving a film which is tougher, harder and flexible. Since the emulsion has already formed a film, this crosslinking and subsequent increase in Tg has no effect upon the film forming properties. The result is a highly crosslinked coating which is optimally insoluble in water and solvents alike, heat resistant and contains no plasticisers or co-solvent.

A number of multifunctional moieties are cited in literature and four have been selected for this presentation.

(a)   Glycidyl Methacrylate (GMA).
(b)   Acetoacetoxyethyl Methacrylate (AAEM).
(c)   The combination of GMA and AAGM.
(d)   Alkoxy Silanes.

Fig 5. Glycidyl Methacrylate

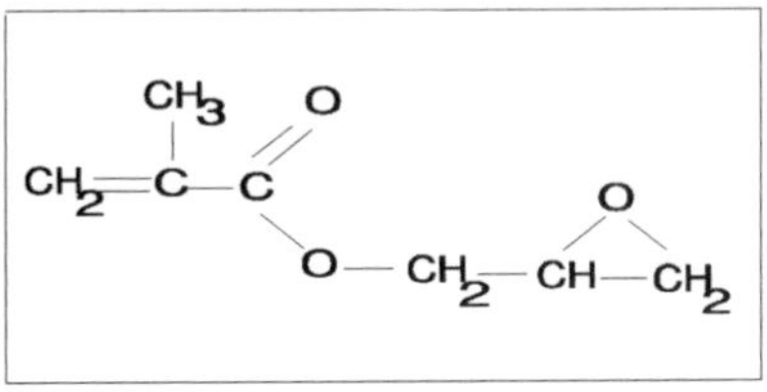

Two reactive sites exist on the molecule.

1)   The methacrylic functionality - normally employed for incorporation of the molecule into the vinyl (acrylic) polymer backbone, leaving pendant epoxy groups.

2)   The epoxy group - this can also be used for incorporation into a polymer by condensation reaction.   Polyethers are formed with pendant methacrylic groups.

Taking the first scenario, the pendant epoxy groups can then be utilised in a number of reactions.   Notably with compounds containing   active   hydrogens   i.e.   amines,   inorganic   acids, carboxylic acids and alcohols.

Derivatives   of   these   reactants   can   thus   be   used   as crosslinking catalysts between epoxy groups i.e. H_2SO_4, diamines, dicarboxylic acids and ethylene glycol.

The epoxy groups can also undergo self condensation to form ether bridges.

The problem with most of these reactions is the requirement for a further polymerisation processing to instigate crosslinking reactions.   The only reactions that will proceed at anything other than a negligible rate at room temperature are the epoxy group with an acidic hydrogen and possibly an amine.

(b)   Acetoacetoxyethyl Methacrylate (AAEM)

This molecule exists mainly as the enol tautomer.

Fig 6. Acetoacetoxyethyl Methacrylate

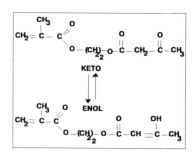

Three reactive sites exist on the molecule.

1)   The methacrylic functionality - normally used for incorporation of acetoacetoxyethyl methacrylate into vinyl (acrylic) polymer chains, whilst leaving pendant acetoacetyl groups available for further reaction.

2)   The methylene hydrogens can undergo numerous reactions with various catalysts, to form crosslinked polymers, such as aldehydes (especially formaldehyde), isocyanates, melamines and electron deficient olefins via the Michael reaction.

3)   The Ketone carbonyl - can either react as the Keto or Enol tautomer.

In the enol state, AAEM will react with diamines at room temperature to produce enamines. Fig. 7

Fig 7. Reaction of AAEM with DIAMINE

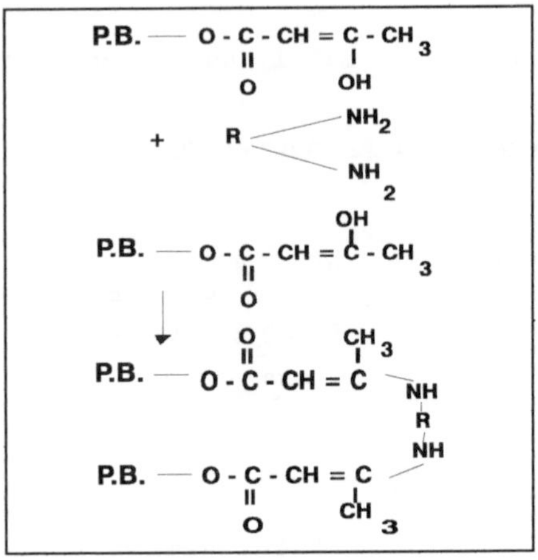

P.B. = Polymer Backbone

All these reactions are basically two pack and in some instances require thermal cure.

(c)   The combination of GMA and AAEM

It has been demonstrated that Glycidyl methacrylate and acetoacetoxyethyl methacrylate may be combined to yield an ambient crosslinking mechanism according to Fig 8.

Fig 8. Reaction of AAEM with GMA

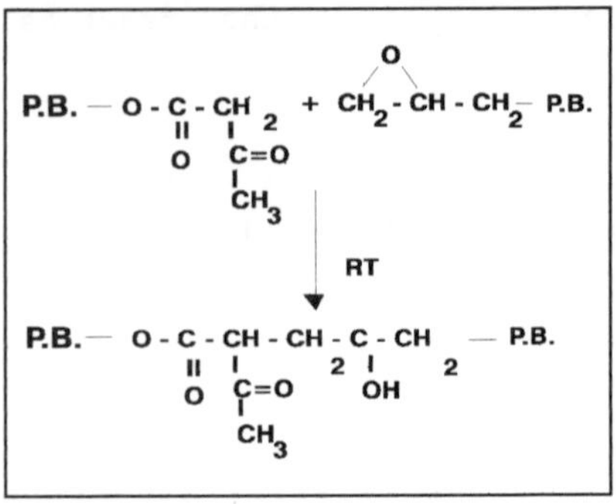

(d)   Alkoxy Silanes

Another effective self-crosslinking mechanism occurring at ambient temperature is to incorporate an alkoxy silane moiety into the copolymer backbone.

The subsequent crosslinking occurs in two stages. Fig. 9

Fig 9. Reactions Of Alkoxy Silanes

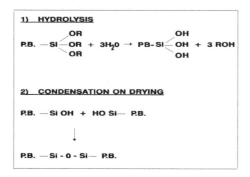

        P.B. = Polymer Backbone

3.3   Performance Characteristics of an Ambient Cure Resin

Over the last two years our own research has led to testing a large number of commercially available latices intended for use in DIY wood lacquer formulations.

Two have been selected for comparison to indicate the improved performance properties achievable by introducing an ambient curing mechanism into the polymer.

Table 9 compares the film properties of two acrylic latices of similar base monomer compositions except for the crosslinking moiety.

The resins were formulated into simple coating systems with water to 35% polymer solids, sufficient coalescent to reduce the MFFT to 0°C, wetting agent and wax.

Glass panels were coated to produce a dry film thickness of 35μ, allowed to cure at 22°C and 55% relative humidity for 14 days before commencing the testing schedule.

Table 9
Performance Characteristics

Resin	A-unmodified	B-ambient cure
MFFT/°C	26	30
Resistance Property		
Dry Heat (100°C)	1	3
Wet Heat (75°C)	3	3
Water	5	5
Coffee	2	5
Vinegar	4	5
Bleach	1	5
Ammonia soln (10%)	1	5
Lemon Juice	5	5
Detergent soln (5%)	1	5
Toilet Spirit	1	4
Potable Spirit	3	5
Acetone	1	4
Ketchup	5	5
Mustard	5	5
Sword Rocker Hardness (secs)	20	22

1 - severe damage

5 - no damage

3.4   **Formulating with an ambient cure resin**

Wood lacquers intended for the DIY market have to meet a number of requirements depending upon the intended applications.

Additives are essential ingredients for maximising performance and as stated earlier judicious selection is imperative.

Two additives have been selected for exemplification.
(a) Coalescing agent
(b) Matting agents

(a) Coalescing Agent

Bearing in mind the requirement to minimize VOCs an efficient coalescent has to be selected. A high dosage of coalescent can effect the early hardness development of films. This is shown in Fig 10. Two have been selected for comparison.

Fig 10.

Effect Of Coalescent On MFFT

Effect Of Coalescent On Film Hardness Development

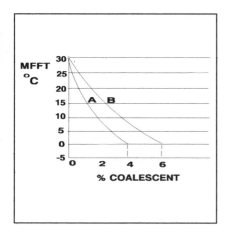

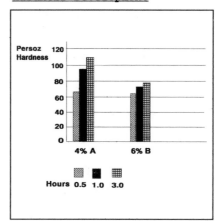

It should also be noted that water miscible coalescents could influence viscosity/rheology control due to the fact that some thickener/rheology modifiers are affected by solvents, particularly the hydrophobically modified products.

(b) Matting Agents

Where lower levels of gloss or sheen are required matting agents are used. There is a variety of products available to the formulator ranging from powdered silicas to powdered waxes and wax dispersions.

Properties such as film clarity can be affected, so once again careful selection has to be considered.

Two have been selected for comparison:-

Fig 11a. Gloss vs Matting Agent Level   **Fig 11b. Film Clarity vs Matting Agent Level**

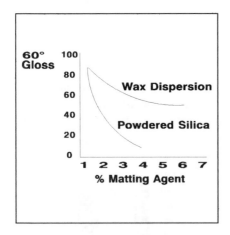   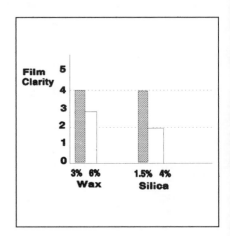

Figures 11a and 11b compares the matting effect of a powdered silica and a wax dispersion and the subsequent effect on film clarity.

e.g. to reduce the gloss by 50%
1.5% silica   - film clarity reduced by 20%
6.0% wax dispersion -   "   "   "   " 40%

CONCLUSION

In the paint industry current research is directed towards the elimination of volatile materials which cause environmental concerns.

For liquid paints, applied by conventional methods, the preferred carrier is water.

To achieve performance characteristics which have been established over a large number of years by solvent based systems, water based resin and additive manufacturers are now providing raw materials which have been well designed for performance in environmentally acceptable systems.

References

1.   J. W. Vanderhoff, E.B. Bradford and W. K. Carrington.
      J. Polym. Sci. Symp., 41 (1973) 155.

2.   S. G. Croll, J. Coat. Technol., 58 (734) (1986) 41.

3.   S.T. Eckersley, Prog. in Coatings, 23(4) (1994) 387.

4.   Dr. Rolf Dersch. Proceedings of 35th Annual Conference
      Surface Coatings Association Australia 1993.

The Preparation and Application of Alkyd Emulsions

G. H. Dekker

ALKYD RESINS DEPARTMENT, DSM RESINS BV, PO BOX 615, CEINTUURBAAN 5, NL-8000AP ZWOLLE, THE NETHERLANDS

1 INTRODUCTION

The use of waterborne alkyds is not new, their history going some 30 years back in time. Their low degree of acceptance obtained in this period suggests that there are still some problems attached to the preparation and use of alkyd emulsions.
    Indeed there are:

* hydrolysis of the backbone, resulting in loss of drying properties upon storage.
* complexation of the cobalt drier with nonionic surfactants of the ethylene oxide type, with neutralizing amines and pigments or fillers, resulting in deteriorated drying properties.
* instability of the paint systems.

Some of these problems, such as hydrolysis of the backbone, can be overcome. The use of hydrolysis resistant raw materials (isophthalic acid instead of orthophthalic anhydride is a well-known example) does not only yield better retention of molecular weight, but also gives better outdoor durability.
Omitting nonionic surfactants with more than 12 ethylene oxide segments as well as omitting amines from the formulation, takes care of the complexation problem.

The instability of paint systems is inherent to the heterogeneity of the system and phenomena like "surfactant theft". The argument that alkyd emulsions are by definition coarse and thus have bad stability, is no longer valid. By careful choice of the surfactant and emulsification equipment, alkyd emulsions can be obtained that have an average particle diameter of less than 500 nm. Nevertheless, careful paint formulation remains a necessity, as is also the case for thermoplastic dispersions.

2 The Basic Theories

To obtain a stable alkyd emulsion, the knowledge of only two formulae is of the utmost importance. For one thing, small particles will have to be made, before being able to protect them from coalescence:

$$d = \frac{\sigma_i}{\dot{\gamma} * \eta_{cont}} * f\left(\frac{\eta_{disp}}{\eta_{cont}}\right)$$

(WU - equation)

For many reasons (stability, gloss, flow, etc.) it is desirable to obtain small particles.
If we consider the Wu - equation, small particles can achieved by:

    A. Decreasing the interfacial tension.
          - emulsifier
          - neutralizing agent
    B. Increasing the shear rate/stress.
          - technology
    C. Decreasing the viscosity of the dispersed phase (alkyd).
          - co-solvent
    D. Increasing the viscosity of the continuous phase (water).
          - thickener

The viscosity of the dispersed/continuous phase also depends on the emulsification process.

    * Direct process   = adding alkyd to water
        - dispersed phase   = alkyd
        - continuous phase   = water
    * Inversion process   = adding water to alkyd
        - dispersed phase   = water
        - continuous phase   = alkyd

Once these particles have been obtained, protection from coalescence can be achieved by ionic as well as by nonionic means. Ionic stabilization is perfectly well described by the well-known DLVO theory and does not need any extensive treatment here. Some observations during experiments that use anionic surfactants will also be treated.

Nonionic stabilization is mainly based upon steric repulsion and osmotic forces between the particles. Specific advantages and disadvantages of this system will be dealt with later.
One should realize that each manufacturer will have his own blend of surfactants, usually determined empirically.

The need for small particles, following Stoke's law, is very obvious:

$$V_{sed} = \frac{(\rho_{cont} - \rho_{disp}) * g * d^2}{18 * \eta_{cont}} \; < \; 1\,mm/day$$

Only in a space shuttle large particle size emulsions can be expected to be stable and this is where reference systems actually have been made.

Where size is concerned, alkyd emulsions have a much larger spread in particle size than acrylic dispersions. Apart from that, also the average particle diameter can be much larger. The better the emulsification process is controlled, the narrower the particle size distribution.

This photograph for instance shows the particle size of an emulsion that has been prepared over the inversion process. It can be clearly seen that, although the inversion had been successful in some regions, it failed in some other, leading to multiple emulsions (water in oil in water). Although very much desired in margarine, such large particles are detrimental to the stability of much lower viscous alkyd emulsions.

When the direct process is used, the alkyd emulsion can have an average particle diameter of less than 1 micron, but then a very low surface tension is needed, usually obtained by anionic surfactants or neutralization with amine. Also the use of excessive amounts of nonionic surfactants (over 5%, based on resin) is an option.

When the inversion process is used and is kept properly under control (this usually means going slowly over the inversion point at very specific empirically determined temperature and solid content),very small particles can be obtained. To obtain sufficient stability a $d_{50\%}$ of less than 750 nm and a $d_{90\%}$ of less than 1000 nm is required. This objective can be reached in the majority of the cases.

3 Comparing Some Parameters

Already some of the parameters determining the overall quality of an alkyd (or indeed any) emulsion have been mentioned, being neutralization and surface tension. Looking at the effect of parameters like these on either particle diminution or stabilization of small particles, it is very helpful to visualize these effects in a table, which is shown below:

Effect of increasing ↓ on →	Particle diminution	Emulsion stability
Viscosity of continuous phase	+ +	+
Sheer rate	+ +	0
Viscosity of dispersed phase	--	0
Temperature	--	-
The amount of:		
Neutralizing agents	+ +	+
Anionic surfactants	+ +	+
Co-solvents	+ +	0
Nonionic surfactants	+	+ +
Thickeners	+	+

We can see that both conditions as well as chemical constituents that are employed during alkyd resin emulsification can serve either one or even both purpose(s). Some typical conditions are: Sheer rate, sheer stress, of course interconnected by the viscosity of the continuous phase, and the temperature. Some typical chemical additives are neutralizing agents (amines, alkali), anionic surfactants, nonionic surfactants (cationic surfactants are hardly used in alkyd emulsification) and co-solvents.

One very interesting conclusion can be drawn from this table: Amine and shear (at least partly) serve the same purpose, i.e. particle diminution. Since the use of amine closely resembles Pandora's box when it comes to problems such as drying deterioration, discoloration of white paints and environmental restrictions, it is advantageous for the paint producer that its use can be minimized by applying the optimal shear, i.e. stirring conditions.

3.1 Main Properties of Nonionic Stabilized Systems

Almost all alkyd emulsions available nowadays contain nonionic surfactants. These have the advantages of being stable in the presence of electrolytes (metal soaps that are used as driers, for instance), although their thermal stability is poor when compared to ionically stabilized emulsions. Since some particle charge is already present as a result of neutralization, anionics are not strictly necessary.

Furthermore, it can be argued that, since absorption of surfactants to particle surfaces is a dynamic equilibrium rather than a static situation, nonionics (being more lipophilic) tend to replace anionics from the surface. These anionics, once "kicked" into the aqueous phase have a very low tendency to re-adhere to the particle. On time-average they are mainly in the aqueous phase, thus giving rise to foaming and flash-rust problems. This was indeed what was observed in practice.

One extremely important aspect of the nonionic stabilized system is the Hydrophilic Lipophilic Balance or HLB.
The HLB, often calculated as the hydrophilic weight fraction of a nonionic surfactant x 20 or determined empirically by titration, determines the time-averaged position of a surfactant with respect to the water-resin interface.

Effect of Changing the Hydrophilic Lipophilic Balance (HLB)

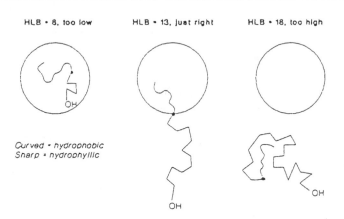

Of course, this phenomenon, like every adsorption isothermal, is dependent upon the concentration of the surfactant in the solution, according to Langmuir.
The choice of the HLB of the surfactant applied will determine your emulsion stability and even your emulsion type. In case of a far too low HLB and a high viscosity resin, the surfactant might never come out of the resin again.
A surfactant of too low HLB will lead to water in oil (w/o) emulsion, just where we want the coating to be o/w, for obvious reasons.
Optimal HLB can be determined by visual observation, although also particle size determination (photo correlation, Mie scattering, Fraunhofer diffraction) can be (and in fact have been) used, which is shown in the next figure:

Particle size versus HLB

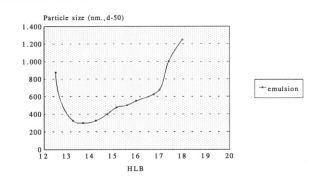

Surprisingly, sometimes even in the case of correctly chosen HLB's (that is, in infinitely stable emulsions), some 25% of the nonionic surfactant can be found in the continuous phase.

It comes as no surprise that the HLB required (RHLB) is largely determined by the polarity of the resin to be emulsified. A high OH-number, for instance, will give rise to a high RHLB. Resins can be modified with respect to OH-numbers by varying the ratio of TMP to pentaerythritol in a standard medium oil soy bean-based alkyd resin.
The higher the OH-number, the higher usually the HLB that was required. This can easily be determined by visual observation. Of course, the same reasoning can be expected to apply to COOH numbers, although here the range is not that large. A similar relationship has been found between RHLB and oil length, oil being a non-polar contribution. Measurements range from pure soy bean oil, over two different alkyd resins, to di-octyl phtalate, being an excellent low-viscous model compound for an oil free polyester.

The results obtained indicate that in fact <u>every</u> resin can be emulsified, regardless of OH-value, oil length and acid number, as long as the HLB is adjusted to these variables. The only <u>real</u> problem we have to deal with when emulsifying alkyd resins is the viscosity of the base resin.

The HLB of a mixture of nonionic surfactants can be influenced to a great extent by the temperature. Each $NP(EO)_n$ has its own cloud point, at which point thermal motion of the ethylene oxide chain causes a decreased hydration. This means a higher Gibbs free energy for the hydrophilic part in water, which in turn means a lower effective HLB.
This is represented in the next figure, showing emulsification of a model resin (RHLB = 13.8) with NP 10 and NP 20 at 5 different temperatures:

Emulsification temperature

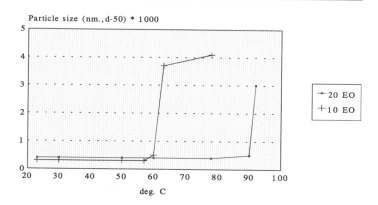

Clearly, NP 10 is more or less useless at temperatures above 55 °C, being its cloud point. Thus a 1:1 mixture of NP 10 and NP 20 (HLB = 14.9) will behave as though it was a smaller amount of NP 20 at an HLB of 15.9 when emulsification is carried out at e.g. 70 °C.

3.2 Anionic, Higher Temperature Systems

As an alternative to the use of nonionic i.e. steric stabilization we can use charge stabilization, brought about by anionic or cationic surfactants, the latter hardly being in use in alkyd emulsification. The advantage of anionic stabilization is the relative insensitivity towards high temperatures, which is of particular importance when emulsifying high viscous resins.

Of the base resin that was mentioned previously, several alkyd emulsions were made using in-line emulsification at elevated temperatures. Here we found a further advantage, being that, while in the case of steric stabilization ± 5 % of surfactant based upon resin is needed, with charge stabilization an amount of 1.5 to 2 % is sufficient.

Interestingly, the not very sophisticated and widely used sodium dodecyl sulphate, one of the lowest priced anionic
surfactants available performed the best. It should be noted however that due to hydrolysis the life time of SDS is limited at higher temperatures, meaning that firstly the residence time in the in-line dispersing unit at high temperatures should be limited to several minutes, and secondly the stability of SDS-emulsions in storage at 50 °C (a test, which is in widespread use in Europe) is bad. Sedimentation as well as coagulation occur after as little as 7 days. Better surfactants are the hydrolysis resistant sulphonates like sodium dodecylbenzene sulphonate.

Concluding it can be said that charge stabilization is clearly far more efficient than steric stabilization.

Earlier it was pointed out that charge and shear stress are more or less interchangeable. Since shear stress is limited by the technical standard of available equipment, the omittance of a neutralizing agent is also limited to lower viscous resins. As a rule-of-thumb we can say that only resins with a viscosity below 5000 dPa.s can be emulsified without neutralization or anionic surfactants. Traditionally, neutralization has been brought about by the use of amines, because they were believed to evaporate during the drying process. This evaporation was necessary to prevent charged loci from staying in the paint film, deteriorating water resistance. This was also the reason why no alkali like NaOH or KOH were used. Clearly Na^+ and K^+ can not be expected to evaporate.

When looking at the weight loss of amine-containing films it can be seen of the most widely used amines (triethylamine TEA, dimethylethanolamine DMEA and aminomethylpropanol AMP) some 25 to 50% actually still is present in the dried film after one day.

The result of this is clearly noticeable when a film, drawn from an emulsion of a 47% soy bean oil alkyd is immersed in water after 1 hr of drying.

When neutralization is accomplished with DMEA (2.1% on resin), the film turns white after 15 minutes. When a high sheer emulsion is made, based on the same resin and sodium hydroxide (0.4% on resin) the same degree of attack by water could be observed after 4 hours.

The effect on water sensitivity caused by improved drying properties on the one hand and less charged loci in the film on the other hand, is clear.

The improvement in drying properties as a result of replacing amines by NaOH can also be visualized by measuring the oxygen uptake, since alkyds dry by reaction of cobalt-generated radicals with oxygen. The more -oxidative- drying, the more radicals and thus the greater the oxygen uptake.

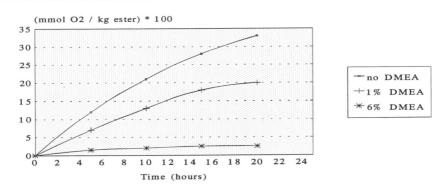

Methyl Linoleate + DMEA
oxygen uptake

One final advantage of this replacement is the reduction of yellowing of white paints. Whereas amine-containing alkyd emulsions tend to give severe yellowing, their NaOH-based counters do not show such behaviour.

4 Mechanical Stability

In cooperation with "Ytkemiska Institutet" we investigated the mechanical stability
of an emulsion. Mechanical stability is an important parameter due to high forces,
to which an emulson is exposed (pumping, grinding, etc.).
From the next graphs we might conclude that the EO-length of a nonionic
emulsifier, as well as the type of the emulsifier (anionic/nonionic), influence the
mechanical stability.

Mechanical stability
EO - length

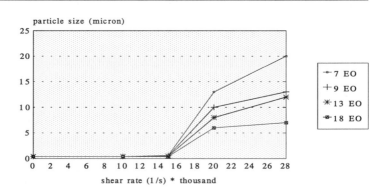

Mechanical stability
Anionic versus Nonionic

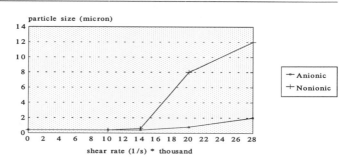

High forces give destabilization of the particles. Lowering the solids content does not affect the phenomenon, see the next graph.

Mechanical stability
Solids content

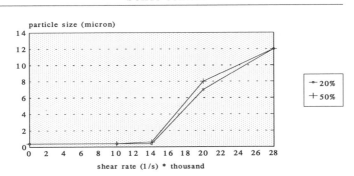

5 Drying Stability

As mentioned before, drying stability is still a problem which has to be overcome. Cobalt, the main drier, should be in the alkyd during the drying stage. Investigation taught us that cobalt will be dissolved in the alkyd when the pH > 7, the next graph.

pH influence on siccative

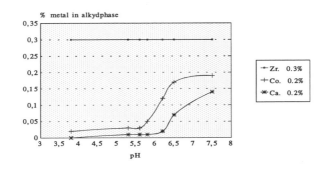

Adding a pigment gave a shift of the cobalt from the alkyd to the pigment after storage of 3 weeks at 50 °C, see the next table.

Location of cobalt (3 weeks 50 °C)		
Emulsion ---------> pH = 7	Alkyd	96%
	Water	4%
Paint --------->	Alkyd	23%
	Water	5%
	Pigment	72%

This phenomenon leads to a loss of drying properties after storage.
However, a first indication shows us that with adding of a pigment it will be possible to influence the shift of the cobalt in a positive way.

6 Conclusion

With the right additives and technology it is possible to emulsify all kinds of alkyds.
Making a paint out of alkyd emulsions still needs some effort, but the available knowledge is growing.

Structure/Property Relationships in Waterborne Epoxy Resin Emulsions

A. Wegmann

POLYMERS DIVISION, BU-RESINS, CIBA-GEIGY AG, CH-4002 BASEL, SWITZERLAND

1 INTRODUCTION

Waterborne 2-component ambient curing epoxy/amine systems are gaining more and more interest in surface protection applications. Their main advantage lies in the almost total absence of organic solvents, therefore causing no fire hazards or air pollution, as well as limiting health, safety and industrial hygiene problems related to organic solvents and vapours.

Advantages in application can also be found, like easy cleaning of the equipment used with water, the excellent adhesion even to substrates normally difficult to coat, e.g. wet concrete, various metals and plastics. However, waterborne compared to solvent-based systems usually have a shorter potlife. Its end cannot be detected easily by visual means, especially in the case of pigmented formulations. The gloss of the cured paint films is strongly dependent on the way the paint is formulated. The drying and curing performance is linked to the evaporation rate of water. Therefore, problems are likely in moist and cold environments. The long-term chemical resistance of waterborne systems, particularly towards acids, is somewhat reduced. The wastewater and the overspray need to be treated in an appropriate manner, otherwise the environmental problem would just be shifted from the air to the water. Despite these limitations, waterborne epoxy systems are nowadays relatively widely used in coatings, predominantly for the following applications:

- protective and/or decorative* coatings for inorganic substrates (concrete, plaster, etc.)
- sealant for freshly poured concrete surfaces to prevent premature drying (mostly in hot countries)
- adhesion promoters, e.g. on wet concrete or for recoating of "old" epoxy coatings
- anticorrosion primers on steel
- primers or single coats* for industrial applications (on metal, curing $\sim$ 80-100°C)
- primers or single coats* on plastic substrates (ambient curing)
- clear coatings* with high gloss on wood

The purpose of this paper is to demonstrate how the type and structure of the epoxy resin influences the coating properties. Resins based on bisphenol A and/or bisphenol F were used. Of course, the curing agent also has a substantial effect. All systems mentioned here have been cured with polyamidoamines (PAA).

---

* non weather resistant

2 EXPERIMENTAL

2.1 Binders

1. Self-emulsifiable liquid epoxy resin based on a mixture of bisphenol A and bisphenol F (®Araldite PY 340-2).
2. Emulsion of an epoxy phenol novolac (EPN) with an average functionality of 2.3 (XB 323; former denomination PF LMB 5123).
3. Emulsion of a solid epoxy resin based on bisphenol A with an epoxy content of about 2.1 eq/kg (XB 320; former denomination PF LMB 5200).
4. Solution of an epoxy resin based on bisphenol A with an epoxy content of about 2.1 eq/kg, 75% in xylene (®Araldite GZ 7071 X-75).
5. Modified water-soluble PAA hardener (Hardener HZ 340).
6. Unmodified PAA hardener for solvent-based coatings (Hardener HY 815-1)

These products and the tradename ®Araldite belong to Ciba-Geigy Ltd., Basel.

2.2 Formulations

2.2.1 Waterborne emulsion systems (see chapter 3.1). Emulsions of bisphenol A or bisphenol F epoxy resins of varying molecular weights were prepared by the inversion method (see chapter 3.2.2) at a solids content of 60%. The same nonionic emulsifier was used at the same concentration. Titanium dioxide RTC 4 (TIOXIDE Ltd., UK) was added to the water-soluble PAA hardener. The paint was formulated by mixing the emulsion and the hardener dispersion.

2.2.2 Waterborne system based on a self-emulsifiable resin (see chapters 3.2-4). 15.4 pbw titanium dioxide RTC 4 were dispersed in 15.4 pbw self-emulsifiable liquid epoxy resin. 18.4 pbw water-soluble PAA hardener were added, the formulation was thoroughly mixed and then diluted with 50.7 pbw deionized water.

2.2.3 Waterborne system based on an EPN emulsion (see chapters 3.2-4). 15.4 pbw titanium dioxide RTC 4 were dispersed in 18.4 pbw water-soluble PAA hardener and diluted with 46.0 pbw deionized water. The paint was then formulated by mixing 20.2 pbw EPN emulsion with the hardener dispersion.

2.2.4 Waterborne system based on a solid resin emulsion (see chapters 3.2-4). 12.7 pbw titanium dioxide RTC 4 were dispersed in 14.3 pbw water-soluble PAA hardener, then 12.7 pbw deionized water were added. The paint was formulated by mixing 60.3 pbw solid resin emulsion with the hardener dispersion.

2.2.5 Solvent-based system (see chapters 3.2-4). 24.4 pbw titanium dioxide RCR 6 (TIOXIDE Ltd., UK) were dispersed in 34.8 pbw of a solid epoxy resin solution (2.1 eq/kg, 75% in xylene). Then 13.0 pbw of unmodified PAA hardener were added followed by approximately 28 pbw of diluting solvents (xylene/n-butanol 4:1) to a viscosity of 70s [DIN 53211, 4mm flow cup, 20°C, 65% relative humidity (rh)].

In all formulations the amount of water or solvent was varied according to the required solids content (% by weight) or viscosity. No additives were added. Pigment/binder ratio: 38.5 : 61.5.

2.3 Application and Test Procedures

Most of the equipment and methods for the preparation and evaluation of the emulsions, as well as for the application and testing of the coatings have been described.[1]

The epoxy content of epoxy resins, solutions or emulsions was measured according to ISO 3001/78. The solids content (% by weight) was determined after heating the sample to 105°C for two hours. Flash points were measured in a closed cup according to ISO 2719/88 or 1523/83. The physical stability factor of an emulsion was determined by subtracting the amount of formed surface water (in %) from the total (100%). The higher this factor, the higher is the stability of the emulsion towards settling or coalescence.

Coatings were tested by the following methods: Cupping test (Erichsen, ISO 1520/73); Impact test (ISO 6272/93); Bend test (cylindrical mandrel, ISO 1519/73); Cross-cut adhesion test (ISO 2409/92); The panels were stored and tested at 20°C, 65% rh. The potlife of waterborne systems was determined by measuring the gloss of paint films applied every 30 minutes after mixing the components. A substantial decline in gloss (>10%) denotes the end of the potlife.

3 RESULTS AND DISCUSSION

3.1 Influence of the Epoxy Resin on the Emulsion and Coating Properties

The relation between the chemical structure of epoxy resins and their performance has been reviewed, with respect to solvent-free and solvent-based but not specifically to waterborne coatings.[2] There is a large variety of different epoxy resin systems for water-based applications on the market. There are emulsions and emulsifiable resins, based on different types of epoxy resins. They contain various types and levels of emulsifiers. Some also contain reactive diluents, organic solvents, other additives, etc. The most dominant factors determining the properties of water-based epoxy resins and coatings are the following:

- chemical structure, molecular weight, epoxy content, and viscosity of the epoxy resin
- amount and type of the emulsifier (chemical structure, HLB-value*)
- amount and type of (possibly) added solvents
- amount and type of (possibly) added additives (defoamers, flow agents, etc.)
- amount and type of curing agent(s)
- physical parameters (emulsion type, particle size, solids content, temperature, humidity, etc.)

It is difficult to seperate the mere influence of the epoxy resin itself from all other influences. Nevertheless, it is of great importance to know what properties can be achieved or altered by means of the epoxy resin. For this purpose, emulsions of bisphenol A or bisphenol F resins of different molecular weights were prepared. The same emulsifier in the same concentration was used, no other additives or solvents were added, and all parameters (manufacturing process, solids content, etc.) were kept as constant as possible.

Figure 1 shows that in the series of bisphenol A resins the potlife is substantially prolonged as the molecular weight increases. The reason for this is the lower number of epoxy groups available for the curing process.

Surprisingly, the same increase of potlife with molecular weight can be observed in the bisphenol F series, where the amount of epoxy groups stays more or less constant (Figure 2). Here the explanation is more likely to be an increased sterical hindrance of the

---

* HLB: Hydrophilic-Lipophilic-Balance[3]

curing process. Because of its high viscosity the highest viscous bisphenol F resin is difficult to emulsify. The poor quality of the emulsion is the reason for the rather short potlife of this system. It is a remarkable fact that an EPN emulsion can be cured with a PAA hardener, whereas in solvent-based systems EPN resins are usually not compatible with PAA hardeners.

In general, it can be observed that bisphenol F resins have a slightly longer potlife than corresponding bisphenol A resins. Bisphenol A resins contain some free hydroxyl groups capable of catalyzing the curing reaction.

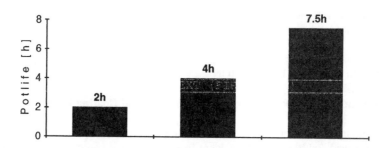

Molecular weight	360	380	500	↗
Viscosity [mPas]	8000	14'000	100'000	↗
Epoxy content	3.2 eq/kg	3.1 eq/kg	2.5 eq/kg	↘

*   DIN 53015 @ 25°C

Figure 1 *Influence of the epoxy resin on the potlife of waterborne formulations (emulsions of bisphenol A resins)*

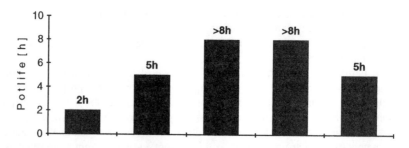

Molecular weight	300	340	380	385	530	↗
Viscosity [mPas]*	1400	6000	40'000	75'000	>100'000	↗
Epoxy content	3.6 eq/kg	3.5 eq/kg	3.3 eq/kg	3.4 eq/kg	3.3 eq/kg	→

*   DIN 53015 @ 25°C

Figure 2 *Influence of the epoxy resin on the potlife of waterborne formulations (emulsions of bisphenol F resins)*

Table 1 *Influence of the epoxy resin on emulsion properties*

Feature \ Epoxy resin	resin based on bisphenol A	resin based on bisphenol F
Molecular weight	low ↗↗↗↗↗↗↗ high	low ↗↗↗↗↗↗↗ high
Viscosity	low ↗↗↗↗↗↗↗ high	low ↗↗↗↗↗↗↗ high
Functionality	c. 2 →→→→→→→ c. 2	c. 2 ↗↗↗↗↗↗↗↗ 3.6
Epoxy content	high ↘↘↘↘↘↘↘ low	high →→→→→ high
Emulsifiability	e a s y | difficult	e a s y | difficult
Physical emulsion stability	poor | g o o d | poor	poor | g o o d | poor
Optimal viscosity range for emulsification (@ 25°C)	10'000 - 100'000 mPas	5'000 - 100'000 mPas
Paint properties	comparable	comparable
Crosslink density	high ↘↘↘↘↘↘↘ low	high →→→→→ high
Required hardener amount	high ↘↘↘↘↘↘↘ low	high →→→→→ high

The emulsifiability of a resin is governed by its viscosity (Table 1). Very high viscous resins (>100'000 mPas) are difficult to emulsify. The physical stability of an emulsion is only good when the viscosity of the epoxy resin lies within a certain range. Very high and very low viscosity resins tend to give unstable emulsions. The paint properties are - with the exception of the potlife - mostly comparable. However, one point has to be carefully kept in mind, that for bisphenol A resins, as the molecular weight increases, the crosslink density and the required amount of hardener decrease. In the bisphenol F series both stay on the same level. Therefore, small differences in flexibility and chemical resistance can be expected.

The viscosity of an emulsion does not depend on the viscosity of the parent epoxy resin (Figure 3). It is rather dependent on the particle size of the emulsion droplets.

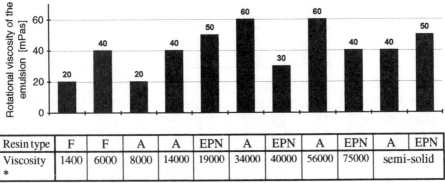

Resin type	F	F	A	A	EPN	A	EPN	A	EPN	A	EPN
Viscosity *	1400	6000	8000	14000	19000	34000	40000	56000	75000	semi-solid	

* viscosity of the epoxy resin [mPas] (DIN 53015, 25°C).

Figure 3 *Viscosity of emulsions (60% solids) as a function of the viscosity of the epoxy resin*

3.2 Waterborne Epoxy Systems Based on Different Resin Types in Comparison to a Solvent-based System

3.2.1 General properties. Based on the finding that epoxy resins of higher molecular weight have advantages, e.g. in terms of potlife, emulsions of EPN and solid bisphenol A resins were developed. Table 2 shows their features in comparison to a self-emulsifiable and to a conventional solvent-based system. The solid resin emulsion needs to contain a small amount of coalescing solvent in order to enable film formation at ambient temperature. This lowers the flashpoint of the emulsion. On the other hand, because of the solid epoxy resin it does not require labelling as "irritant" in Europe.

3.2.2 Particle size of emulsions. Figure 4 shows the particle size distribution of the EPN- and the solid resin emulsions. Both have a mean particle size of well below 1µm. The particle size of an emulsion can be influenced by the amount and type (chemical structure and HLB-value) of the emulsifier. High amounts of very hydrophilic emulsifiers tend to give emulsions with low particle size. Which chemical structure is best suited for a certain resin has to be found out empirically. The manufacturing equipment and process also play a very important part in getting a low particle size. Epoxy resins of medium viscosity are best emulsified using a high speed disperser and the so called "inversion method".[4] Water is added slowly to the resin until a phase inversion from a water-in-oil (w/o) to an oil-in-water (o/w) emulsion occurs. At this stage high shear forces are used to break down the emulsion droplets to a small size (Figure 5).

Table 3 shows why particle size is such a crucial property. The most important point is that emulsions with small particle size have a better curing performance. This leads to coatings with better gloss, hardness and resistance to chemicals. An explanation for this effect has been proposed.[1]

Emulsions with low particle size also have a higher viscosity and, therefore, a higher physical stability, i.e. they have a reduced tendency to settle or to coalesce. Figure 6 shows that over a period of 1 year both epoxy resin emulsions are very stable at 20°C.

Table 2 *Waterborne compared to solvent-based epoxy resins*				
Product Feature	waterborne self-emulsifiable epoxy resin	waterborne EPN emulsion	waterborne solid resin emulsion	solvent-based epoxy resin
Resin type	bisphenol A/F liquid resin	EPN resin	bisphenol A solid resin	bisphenol A solid resin
Solids content	100%	76%	55%	75%
Emulsifier (external)	nonionic	nonionic	nonionic	none
Reactive diluent	0%	0%	0%	0%
Organic solvent	0%	0%	9%	25%
Flash point	> 200°C	> 200°C	42 - 43°C	26°C
Epoxy content [eq/kg]	5.5 - 5.8	4.0 - 4.5	1.01 - 1.13	1.50 - 1.67
Labelling (Europe)	irritant	irritant	---	harmful,
	---	---	flammable	flammable

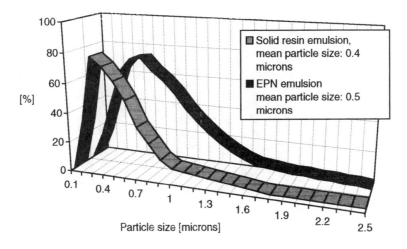

Figure 4 *Particle size distribution of epoxy resin emulsions*

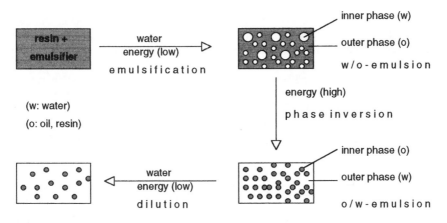

Figure 5 *Emulsification process (inversion method)*

Table 3 *Effect of particle size on the quality of the emulsion and the coating*		
lower (< 2µm) <------------------- particle size ------------------->	higher (> 3µm)	
higher <------------------------------ viscosity ------------------------------>	lower	
higher <------------------------------ physical stability ------------------------------>	lower	
complete, <--------------------------- curing of the -------------------> incomplete,		
homogeneous	coating	inhomogeneous
higher <------------------------------ gloss ------------------------------>	lower	
higher <------------------------------ hardness ------------------------------>	lower	
higher <------------------------- chemical resistance -------------------------->	lower	

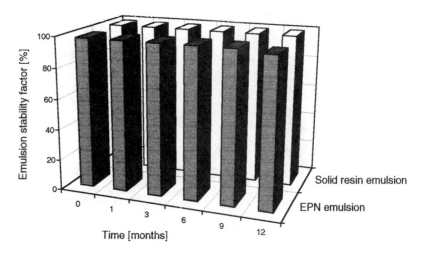

Figure 6 *Physical stability of epoxy resin emulsions at 20°C*

3.2.3 Potlife. The gloss of a solvent-based system is completely stable throughout its potlife. This is not the case for waterborne systems, therefore the decline in gloss can be used to determine the end of their potlife in pigmented formulations (Figure 7). The potlife of a self-emulsifiable resin is relatively short when it is formulated in a 3-component system, i.e. the water is added at the end in a seperate step. In a 2-component system, where the water is added in advance to either the resin or the hardener component, the potlife is even shorter. The EPN- and solid resin emulsions, on the other hand, have much longer potlives, independent of the way they were formulated. This is a big advantage because for practical reasons formulators strongly prefer 2- over 3-component systems.

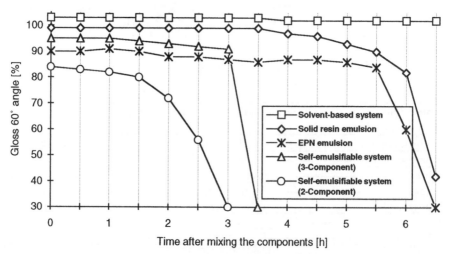

Figure 7 *Gloss stability of water- and solvent-based systems during the potlife*

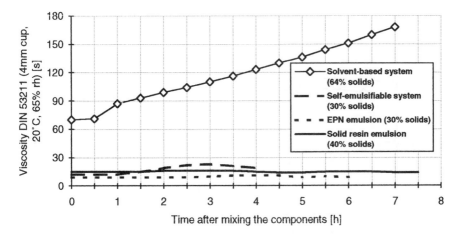

Figure 8 *Viscosity of pigmented epoxy formulations during the potlife*

In a solvent-based system the beginning reaction between resin and hardener leads to an increase in viscosity which makes the end of potlife "visible". In water-based systems this reaction and viscosity increase takes place predominantly inside the emulsion droplets. Therefore, it does not contribute to the measurable external viscosity of the system. In this case the end of potlife may not be directly visible, particularly in pigmented formulations (Figure 8).

Although measured by different methods, the duration of the potlife of EPN or solid resin emulsions equals that of a comparable solvent-based system.

3.2.4 Viscosity. Figure 9 shows that the shear viscosity of a liquid resin or a solid resin solution is independent of the applied shear rate (Newtonian behaviour). The EPN emulsion is strongly dependent on the shear rate (pseudoplastic or non-Newtonian behaviour). The solid resin emulsion is also influenced by the shear rate, but its non-Newtonian character is less pronounced. This is most likely due to the small amount of organic solvent it contains.

Regarding formulated paints, a similar behaviour is observed (Figure 10). The solvent-based system is Newtonian, the waterborne systems are pseudoplastic. Again the solid resin emulsion is less pseudoplastic than the other waterborne systems because of the organic solvent content. At high shear rates the viscosities of all systems are comparable. But under low shear stress the waterborne systems have a much higher viscosity than a solvent-based system.

3.2.5 Paint properties. Table 4 indicates that waterborne systems have to be applied at a considerably lower solids content compared to the solvent-based. An exception is the solid resin emulsion, due to the small amount of organic solvent it contains. Because of the pseudoplastic behaviour the emulsions can be applied easier than the liquid resin. The high shear forces generated in grinding pigments make it difficult to pigment emulsions. Thus, it is advantageous to incorporate pigments and fillers into the hardener. The solvent-based system and the solid resin emulsion dry faster than the other two waterborne systems due to the physical drying of the solid epoxy resin. The waterborne systems have a much lower impact resistance than the solvent-based, and also a slightly lower resistance to boiling

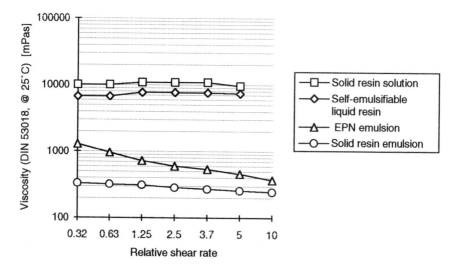

Figure 9 *Rotational viscosity of epoxy resin components*

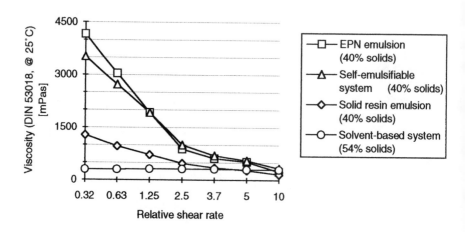

Figure 10 *Rotational viscosity of unpigmented paint formulations*

water. The solid epoxy resin emulsion cures a little slower than the other systems and has a slightly reduced acetone resistance. However, with the exception of the lower impact resistance the performance of the waterborne emulsions comes very close to that of the solvent-based system.

Table 4 *Paint properties of waterborne in comparison to solvent-based coatings*

Product / Feature	waterborne self-emulsifiable epoxy resin	waterborne EPN emulsion	waterborne solid resin emulsion	solvent-based epoxy resin
Diluents	water	water	water	xylene/ n-butanol
Solids content (application)	40%	40%	53%	64%
Resin:hardener				
tel quel	100 : 120	100 : 90	100 : 24	100 : 38
solid/solid	100 : 60	100 : 59	100 : 23	100 : 50
Pigmentation	resin	hardener	hardener	resin
Dust-dry time[1]	5 - 6h	5 - 6h	3 - 4h	3 - 4h
Full cure time[1]	11 - 12h	11 - 12h	12 - 13h	9h
Film aspect[1]	ok, high gloss	ok, high gloss	ok, high gloss	ok, high gloss
Persoz hardness				
after 1 day[1]	220s	220s	150s	180s
after 7 days[1]	320s	300s	220s	300s
after 14 days[1]	350s	325s	280s	310s
Cupping test[2] (Erichsen)	8 mm	9 mm	8 mm	9 - 10 mm
Direct impact[2]	40 - 60 kg cm	40 - 60 kg cm	20 - 40 kg cm	>160 kg cm[3]
Bend test[2] (cyl. mandrel)	3 - 5 mm	3 - 5 mm	3 - 5 mm	1 mm
Cross-cut test [2] (adhesion)	excellent	excellent	excellent	excellent
Acetone rub test (20x ↔)[2]	unaffected	unaffected	scratchable	unaffected
Boiling water resistance (6h)[2]	slight blistering	slight blistering scratchable	slight blistering scratchable	unaffected

[1] on glass plates
[2] on pickled, cold-rolled steel panels 14.03
   curing time: 14 days
[3] reverse impact

Curing: 20°C, 65% rh
Dry-film thickness: c. 30 μm
Pigment: titanium dioxide

3.2.6 Chemical resistance. The long-term chemical resistance of waterborne compared to a solvent-based system is reduced, especially towards acids (Table 5). Compared to the self-emulsifiable system, the EPN emulsion shows a slightly improved resistance towards ethanol, whereas the solid resin emulsion gives a little better resistance to lactic acid and ammonia. Short term exposure of waterborne coatings to these chemicals (spills), however, should provide no problem at all.

Table 5 *Chemical resistance of waterborne in comparison to solvent-based systems*				
Product Test reagent	waterborne self-emulsifiable epoxy resin	waterborne EPN emulsion	waterborne solid resin emulsion	solvent-based epoxy resin
Deionized water	■■■■■■■■■	■■■■■■■■	■■■■■■■■	■■■■■■■■
Sodium chloride 10%	■■■■■■■■■	■■■■■■■■	■■■■■■■■	■■■■■■■■
Hydrochloric acid 10%	Destroyed	Destroyed	Destroyed	■■■■■■■■
Sulfuric acid 10%	Destroyed	Destroyed	Destroyed	■■■■■■■■
Lactic acid 2%	Destroyed	Destroyed	‖‖ Destroyed	■■■■■■■■
Sodium hydroxide 50%	■■■■■■■■■	■■■■■■■■	■■■■■■■■	■■■■■■■■
Ammonia 10%	‖‖‖‖‖‖‖‖‖‖‖‖	‖‖‖‖‖‖‖‖‖‖‖	■■ ‖‖‖‖‖‖‖‖‖	■■■■■■■■
Xylene	■■■■■■■■■	■■■■■■■■	■■■■■■■■	■■■■■■■■
Ethanol 95%	■ ‖‖‖‖‖‖‖‖‖‖	■■■ ‖‖‖‖‖‖‖	■■■ ‖‖‖‖‖‖‖	■■■■■■■■
Ethanol 30%	■■■■■■■■■	■■■■■■■■	■■■■■■■■	■■■■■■■■
Test duration [months]	¼ 1 3 6 9 12	¼ 1 3 6 9 12	¼ 1 3 6 9 12	¼ 1 3 6 9 12

Substrate: pickled, cold-rolled steel panels 14.03   Immersion temperature: 20-22°C
Application: spraying   ■■■■   unaffected
Dry-film thickness: 100 ± 10 μm (3 layers)   ‖‖‖‖   attacked
Curing: 14 days, 20°C, 65% rh

4 CONCLUSION

Table 6 summarizes advantages (upper part) and disadvantages (lower part) of the tested waterborne and solvent-based coating systems. The big advantage of the waterborne self-emulsifiable system are the environmental and safety aspects, whereas the solvent-based systems have better application properties. By using waterborne emulsions of high viscosity or solid epoxy resins, based on bisphenol A or F, it is possible to considerably improve application and end properties of the coatings while maintaining the environmental benefits. The largest remaining gap between aqueous and solvent-based systems remains in the long-term acid resistance.

Table 6 *Summary of properties*

waterborne self-emulsifiable resin	waterborne EPN emulsion	waterborne solid resin emulsion	solvent-based epoxy resin
environmental and safety benefits			high impact
excellent on wet concrete and plastics, interlayer adhesion			good flow
no storage problem	pseudoplastic	pseudoplastic	no storage problem
2-pack system unfavourable	2-pack system ok	2-pack system ok	2-pack system ok
short potlife	long potlife	long potlife	long potlife
slow drying	slow drying	fast drying	fast drying
high viscosity	low freeze/thaw stability	moderate freeze/thaw stability	good chemical resistance
moderate flow	moderate flow	moderate flow	high viscosity
low impact	low impact	low impact	unsuitable on wet concrete or plastics
low chemical resistance	moderate chemical resistance	moderate chemical resistance	environmental and safety hazards

Acknowledgements. The author thanks his coworkers Ch. Studer, O. Huynh and R. Cioffi for carrying out the experimental work, and J-M. Pfefferlé for particle size analysis of the emulsions.

References

1.   A. Wegmann, *Journal of Coatings Technology*, 1993, **65** (No. 827), 27.
2.   D.L. Steele, *Surface Coatings Australia*, 1992, **29** (No. 10), 6.
3.   R.C. Sonntag, *Journal of Coatings Technology*, 1988, **60** (No. 757), 53.
4.   G.Dahms, O. Hafner, *Paint & Resins*, 1988, No. **8**, 13.

Aqueous Polymeric Coatings for Textiles

T. Matthews

DEVELOPMENT MANAGER, MYDRIN LIMITED, CARLTON INDUSTRIAL ESTATE, BARNSLEY
S71 3PL, UK

1 INTRODUCTION

Textiles have been coated in some form for at least 1000 years. Stiffening, fabric stabilisation and water proofing are some of the more well known applications of coated textiles.

Until relatively recently, ie. since World War I, coating materials have been based on naturally occurring products such as starch, bitumen, tars and natural rubber. The rubber naturally occurs as aqueous latex but, the rubber was usually recovered as 'crepe' and when used for coating was applied from solution.

The advent of new synthetic polymers around the 1930's and 40's resulted in new fibres and textiles appearing creating new markets and opportunities. Coatings provided a straight forward way for many fabric finishers to modify existing fabric stock to meet the new requirements. The development of synthetic rubber (styrene-butadiene) gave impetus to the development of many other polymers manufactured by the emulsion polymerisation method, providing the coater and formulator with many more options.

Coatings can be applied to textiles in solid form eg. using 'hot melt' systems, or using liquid cross-linkable products, but here use is restricted to 'high build' fusibles only suitable for limited applications eg. PVC plastisols for tarpaulins.

Solutions of polymers in organic solvents are used in some areas, eg. proofed nylon works well but have many problems:

    a) solvent removal and recovery
    b) low solids necessitating in many cases multiple coat applications
    c) viscosity control and modification is difficult because of the strong contribution of the base polymer.

By far the largest use is for aqueous polymer dispersions produced by the emulsion polymerisation process.

The mechanism of this process ie. polymerisation within surfactant micelles distributed throughout the aqueous medium results, in general terms , in:
  a) high molecular weights.
  b) low viscosities.
  c) high solids content.
  d) no problems with water removal. Textile processing is usually carried out at temperatures > 100°C and often at 140°C-150°C.

These properties enable the formulator to modify and control the dry film properties and simultaneously the aqueous phase. Viscosity, rheology can be varied considerably, and if foaming is a requirement, then surfactants can be included to control both the fineness and the stability of the foam structure.
These properties together with the versatility of aqueous systems make them the natural choice for the great majority of coating applications.

1.1 Principles

Coated textiles are essentially a composite structure where three factors affect the final result, ie.:
1.1.1 The textile itself: the construction can be woven, knitted, non woven. Simple or mixed yarn types can be used and for pile fabrics (velours) the pile yarn can be different from the backing yarns. The variety here is considerable and for optimum results, each fabric type needs to be individually assessed.
1.1.2 The coating: based on emulsion polymers and other additives to control the dry film properties dictated by the end use.
1.1.3 Method of application: this brings textile and coating together and affects the position of the coating on, or within the textile structure. This controls important textile properties eg. handle or feel.
1.1.4 In summary then, the objective is to:
a) identify the performance parameters required.
b) design the dry film characteristics to meet these
   requirements.
c) control the aqueous phase to ensure that the coating can be
   properly applied to the textile.
1.1.5 It is possible to design coating systems to meet several technical parameters simultaneously. Alternatively one fabric type can be converted to meet the requirements of more than one market place by the design of different coating products.

There are many methods of coating textiles:
  1. Knife or spread coating using paste
  2. Knife or spread coating using foam
  3. Screen printing
  4. Transfer coating
  5. Impregnation
  6. Spraying
  7. Lick roll

In reality, in dealing with one particular end use, only one system will be available and the design elements must include both the finished textile properties and the appropriate aqueous phase requirements in terms of eg.

(i)   Viscosity
(ii) Rheology
(iii) Foaming properties where appropriate

For the polymer chemist then we have quite a complex scenario:

a) Evaluation of the textile
   Construction
   Yarn type(s)
   desired end use parameters.
b) coating equipment.
c) Selection of polymer(s) and other additives.
   Where, for example the finished goods are required to be durable to washing and dry cleaning, then cross-linkable polymers must be used. In some cases more specialised polymers eg. high nitrile type may have to be used to meet high performance dry cleaning specifications.

Health & Safety aspects have now become very far reaching and many previously usable chemicals and additives are being ruled out on these grounds.

Some examples of where the various polymer types find their major use is shown below and also serves to illustrate the variety of applications within the textile industry.

PRODUCT	POLYMER	REASON FOR USE
Carpets	SBR	Tuft lock, handle
Rugs	SBR/N.R.	Tuft lock, handle/wash
Bath mats	SBR/N.R.	Tuft lock, handle/wash
Knitted	Acrylics, VA/E	Stability,wash & d/clean
Wovens	PVC's, acrylics	FR, seam slip
Velours	PVC's, acrylics	FR, tuft lock
Automotive	PVC/VA/E	FR, H.F. welding, wear
Louvre	PVA, PVC	Stiffening, FR welding
Filter cloths	PU's, acrylics	Pore size, easy clean
Interlinings	PE's & PA's	Hot melt adhesives
Wash Leather	Nitrile/SBR's	Wash, abrasion
Glass	PVA's,SBR, Acrylic	Seam slip, Alkali resistance
Drapes	Acrylics, PVC's	Thermal linings, FR
Mattress ticking	Arylics, VA/E's	Seam slip, abrasion
Ironing boards	Acrylics	Heat & rub resistance, wear

The base polymer is normally compounded with many other additives to achieve the final result. These are typically:
1. Fillers eg. clays, silicates, carbonates to modify handle and give cost reduction.

2. Plasticizers: conventional or FR types eg.
   phosphates.
3. Waxes for water repellancy, slip.
4. Viscosity, rheology modifiers.
5. Surfactants both for dispersing and wetting of eg.
   pigments and fillers and for foam control, either
   during manufacture or at the application stage.
6. Fire retardants eg. Bromine/Antimony oxides.

The range of technical options available to the chemist
is considerable, although in fact many elements of the design
requirements serve to eliminate many of these quite
quickly.

The final stage is to take the fully formulated coating
product to the coating equipment and produce coated fabrics.
Further refinement of the product may be necessary to optimise
the end result, many of these changes will be carried out 'on
site' with the user to ensure fast response and feed back.

1.2 Case Studies

1.2.1 The Domestic Environment. Coated fabrics play a
major role in the house. Carpets are back coated usually with
SBR products to provide stability and for some constructions
like tufted carpets the coating anchors the tufts, it can also
function as an adhesive when a secondary backing fabric is
applied to provide additional wear properties.

Other examples are found in furniture fabrics where back
coatings confer wear properties, seam slippage and fire
retardant performance to meet the current UK
regulations.[1]

Mattress tickings are coated primarily to give seam[2]
slippage, louvre blinds are stiffened with weldable
properties to make the pocket at the bottom of the louvre to
take the hanging weight, wallcoverings also use coatings for
easy clean finishes, lamination and in producing textured
finishes with chemically or physically 'blown' coatings.

Finally, curtains and drapes utilise coating technology
enabling the conversion of relatively low grade fabrics into
aesthetically acceptable high value products. This
application will now be discussed in detail.

1.2.1.1 Drapes. Curtains and drapes combine aesthetic
and functional properties, ie. they provide an effective
thermal barrier. Curtains are frequently lined, either 'sewn
in' or as a separate hanging

(a) to protect the primary curtain, and
(b) to add to the insulative properties.

Aqueous coatings, using the so called 'crushed foam'
route have enabled the production of high value drapes/linings
from low cost textiles. The coating provides added bulk,
aesthetics and thermal insulation properties to an otherwise
unsuitable fabric.

In principle the process is straight forward ie.
typically a fully formulated acrylic emulsion polymer is
foamed to a low S.G. (0.2) applied as a positive layer (~0.5 mm)

to the fabric, flocked and dried at <100°C. It is then crushed through nip rollers or a callendar, and finally baked at approximately 140°C to fully cross link the system, maximising adhesion and enhancing properties like washing and dry cleaning.[3]

This process demonstrates the composite nature of coated fabrics and the significance of how and where the coating is placed as well as the actual chemical composition of the coating itself. In more detail then we have:

1. Polymer: Acrylic polymers provide an ideal starting point for many textile applications, they provide a pleasant round handle to the fabric, and acrylic monomers copolymerise readily, both with other acrylates, methacrylates and monomers like styrene, acrylonitrile etc., providing additional properties such as improved water and/or solvent resistance.

Softness is an important factor for textile coatings, particularly for curtainings and glass transition temperatures are normally less than 0°C and frequently -20 to -30°C.

Copolymers of ethyl and butylacrylates are typical and to meet the durability requirements of most textiles means that cross linkable grades are used either carboxylated or copolymerised with N-methylol acrylamide. In many applications the additional use of small amounts of eg. melamine formaldehyde resins 'tightens' the cross link structure and for some yarns eg. polyesters shows benefits in terms of adhesion.

2. Foaming. The advantage of foam is that a softer finish is produced when compared with unfoamed, paste products. Foams come in many forms however, eg. foam from the very coarse, unstable foams typified by washing up liquids through to very fine stable foams eg. aerosol shaving foams or meringue. It is this latter type which is relevant to the curtain liner application, in that a layer of low density wet foam on the textile will penetrate into the structure only a relatively small way. Excessive penetration leads to stiffening and loss of the required drape character.

An additional factor is that the foam structure must be heat stable and not show problems such as crazing, mud cracking and delamination at the coating/textile interface.

The design therefore of the surfactant system is critical and all the parameters cannot be achieved with one type. Typically a primary, heat stable foaming agent is used coupled with additional surfactants to refine the foam structure.

The rheology of this foam is also crucial in that the knife coating process imparts high shear to the foam surface particularly at speeds of 20 metres/minute or more.

3. Flocking. Normally immediately after coating flock, either ground or short staple (eg. 2-3mm) flock is fired into the wet adhesive under an electrostatic charge, this aligns the fibres vertically which improves penetration in to the wet foam. The flock improves the feel of the finished fabric giving an impression of quality, and in addition abrasion properties are improved.

4. Drying and Crushing. This stage is the key step in the process. The foam before crushing is weak and even if fully cured would be readily removed by normal handling. Crushing the foam in to the fabric structure provides the following benefits:
   a. increases adhesion
   b. reduces abrasion problems
   c. retains the soft drape character
   d. some foam structure is retained thereby ensuring a breathable fabric.

It is essential that the foam is only dried ($<100\,°C$) before the crushing process. Higher temperatures will initiate the cross-linking and the foam will recover, returning to its original dimensions and the benefits of this process will be lost.

5. Curing. The final stage is to fully cross-link the polymeric binder and temperatures of ~$140\,°C$ are required for this.

6. General Processing. The textiles are normally coated continuously, fabric rolls lengths varying from a few hundred metres to several thousand metres. The aqueous acrylic is mechanically foamed through a continuous mixer. The compound is pumped into a mixing head where air is introduced in a controlled fashion to produce foam of the desired density and quality.
The coating operation is normally by vertical knife over roller, where the knife is positioned over the heart of a large roller. The gap between the knife and fabric is adjustable depending upon the final coating weight required, 0.5 mm is typical. Delivery of foam to the bank is via a pipe continuously traversing across the foam bank behind the knife.
Flocking is carried out directly onto the textile in a fully enclosed unit. The coated fabric is held under tension on stenter pins set at a pre-determined width and carried through the drier at speeds of 20-60 metres/minute dependent upon the efficiency of the unit.
Crushing is usually through nip rollers chromed on the coated side, rubber on the fabric face. Pressures are relatively low eg. 5-6 bar. Final cure requires fabric temperatures above $120\,°C$, ideally $140\,°C$, actual set temperatures will be higher than this depending upon the running speed and residence time in the oven.

A schematic of line set up is as per Fig. 1 attached.

Multiple coat finishes can be handled by this route, providing that the cure step is left until the final coat, this maximises intercoat adhesion. Flocking between coats is not usual, but ground floc is sometimes used to minimise blocking in the rolled fabric.
Blackout curtaining is produced this way and is either two coat sometimes called "dim out" or 3 coat. For full black out the sequence of coats are white – black – white to maximise

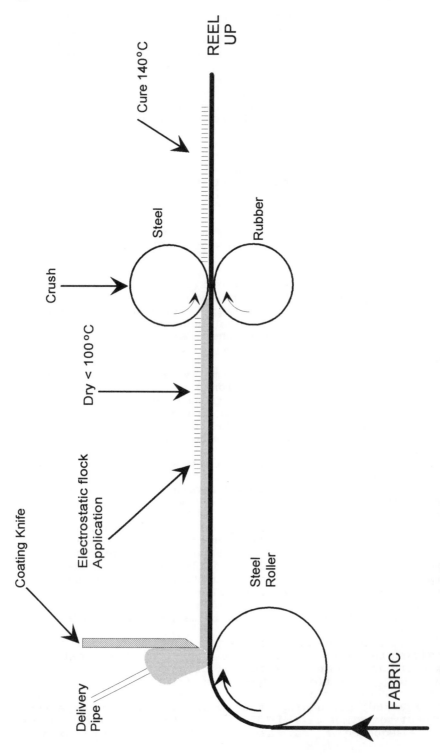

Figure 1 *Schematic of crushed foam coating process for curtains and drapes (not to scale)*

opacity. The final coat can be other colours, cream or ecru are common.

The products are evaluated for wash performance, usually at 40°C (HLCC6)[3], and dry clean[3] generally using commercial units, although equipment complying with this standard is available through test houses for a more formal evaluation.

1.2.2 The Industrial Environment. Functional properties predominate here, unlike the domestic situations where aesthetics are an essential factor.
Some examples incude:

1. Filter fabrics: foam coating provides porosity control and easy cleaning properties. Frequently hostile environments are encountered eg. hot moist air at 100-20°C and/or a range of chemical vapours.
2. Protective clothing for oil rigs, fire fighting and riot suits. Here, heat reflectivity is required and up to 200°C is common, together with resistance to oil, petrol and chemicals. Yarns are often specialised and include carbon fibre and aramids.
3. The building industry uses glass fibre fabrics for a wide range of uses and this particlar application will be discussed in more detail.

1.2.2.1 Glass Fibre Textiles Woven and non-wovens are used for a wide range of applications eg.

1. Reinforcing scrims, for cement rendering.
2. Acoustic ceiling tiles.
3. Wall coverings for hospitals and similar environments.
4. Glass wool laminated to paper or aluminised paper.

Glass fibre has excellent tensile strength coupled with minimal extension and is non-burning, ideal properties for many applications.

From the textile viewpoint glass fibre has two major problems, the fibres are:

a) Slippy, and therefore eg. woven constructions are very unstable.
b) Brittle, resulting in loss of tensile properties and also presenting handling problems.

Aqueous coatings address these problems effectively and additionally in hostile alkaline environments, eg. cement, protect the glass from attack.

1.2.2.1.1 Building Scrims. This application illustrates all the above situations quite well. These scrims are used to reinforce cement rendering and pre-cast structures.

The construction of the fabrics are an open mesh design, typically having a mesh size of 0.5 cm square.

A coating has to stabilise the construction for subsequent handling. Without the coating the fabric would literally fall apart. In addition to the stabilisation the finished textile needs to have a certain level of stiffness again for ease of subsequent handling.

Finally the coating must protect the yarns from the aggressive environment found in cement mixtures ie. caustic alkali for extended periods.

The coating material needs to fulfil several requirements:

1. stabilisation : ie to hold the fabric together and prevent break up of the brittle fibres which can create dermatitic problems.
2. stiffening : to improve handling characteristics.
3. alkali resistance : within cementitious screeds the concentration of caustic alkali is sufficient to attack the glass.

Performance of the finished fabric is measured against National standards of which the Austrian Standard O Norm 6122 is typical.[4]

The choice of polymer type(s) is crucial to achieve these requirements and typically acrylics using melamine formaldehyde resins to maximise cross linking are used, or alternatively and increasingly styrene-butadiene based formulations are used primarily because of their better alkali resistance and freedom from formaldehyde.

Application to the glass scrims are traditionally by simple padding techniques although where available screen printing technology is being used to improve production speeds and thereby cut production costs.

1.2.2.1.2   Wallcoverings.   Woven glass fibre wallcoverings are used widely in hospitals, offices and similar institutions.

The properties required in the finished glass are considerable ie.

1. Weave lock, particularly when wet ie. with wallpaper paste.
2. Infil of the weave structure.
3. Flexibility such that the fabric can readily accommodate corners, bends etc.
4. Ability to accept paint; a good coating reduces the paint usage considerably.
5. Fire retardant, normally to the German DIN Standard 4102.[5]

Traditional coating systems are starch based and use zirconium salts to provide the necessary water resistance. Fire retardancy is by a halogen/antimony synergistic system. This system works well but environmental and ecological pressures are demanding newer, friendlier products and development work is in hand with new systems which are:

a) all synthetic polymers
b) are completely halogen and antimony free.

The method of application is by double sided screen: "Screen to Screen" a relatively new technique.

Progress to date with this new product has been excellent and full scale production should commence by the end of the year.

This development resulted from a complete re-assessment of what we were trying to achieve and the end result combines:

a) an emulsion polymer (non V.C.) which combines

flexibility and contributes to the fire retardant
performance.
b) a soluble synthetic polymer which aids film
formation and controls viscosity.
c) a non halogenated inorganic fire retardant
combination providing the necessary fire
retardancy, low smoke without contributing to the
toxic loading.
The final product now meets all the major technical
requirements and represents a significant step forward.

1.2.2.1.3 Acoustic Ceiling Tiles. These are made from a
non woven web where the coating again fulfils several
functions:
1. Decorative
2. Dust/fibre control
3. Fire retardancy
4. Stiffening
The composition of products for this area is close to emulsion
paint technology: stiffer (high Tg) polymers are used, usually
vinyl acetate copolymers and the filler systems normally
include relatively high levels of Titanium dioxide for
capacity and whiteners.

An interesting additional area here is in the production
of mineral wool ceiling tiles where a glass non woven fleece is
laminated with an aqueous fire retardant adhesive, normally
spray applied. The finished composite is then painted with a
fire retardant decorative paint.

1.2.2.1.4 Glass Wool Roof Insulation. This product is
made by extruding molten glass directly on to carrier belts
where phenolic resins are used to bond the fibres. At a later
stage either paper or aluminised paper is laminated to the
glass webb. Aqueous vinylidene dichloride copolymers are used
to maintain the necessary fire retardancy (DIN4102) and
provide the essential hot tack characteristics to achieve
satisfactory bond strength.

1.3 Future Trends

1.3.1. Environmental and ecological pressures provide
the major short to medium term drive for change. Aqueous
coatings whilst well established in the textile market
represent a significant step for many other coating areas.

Solvent coatings are still used in some areas of textile
finishing eg. polyurethane waterproofing and solvent
recycling and reuse is much better, particularly post
COSHH,[6] the developments here are to produce aqueous
versions having the same performance.

Breathable, or more accurately hydrophilic polyurethanes
are becoming available which allow moisture transmission
through the coating allowing increased comfort levels in
garments without incurring the cost penalties of more expensive
laminated structures.

1.3.2. Fire retardancy is a very important area for
textiles, specifically furnishing fabrics where, for example,
there is legislation in place in the U.K. This demands that for
domestic furniture the covering fabric is resistant to smokers'

materials ie. cigarette and match. The most universal fire
retardant chemistry available is based on halogen/antimony
oxide both of which are now under pressure on health grounds.
The most widely used bromine donor is deca-bromodiphenyl oxide
which can under certain exceptional cicumstances produce
dioxin derivatives. Antimony trioxide has been classified as
a suspected carcinogen.

The pressure to ban or restrict the use of these products
is considerable but the risk from these chemicals is arguable,
much smaller than risk of fire which they actively reduce.
Current alternative chemistry eg. phosphorus derivatives have
there own hazards as yet not in the limelight.

What is required is a proper balance assessment of risk
versus benefit of both existing and alternative technologies
with a sensible time interval for better systems to be
developed, if necessary.

1.3.3. Recycling and recyclable polymers and
composites are becoming increasingly important as waste
disposable reaches ever larger proportion.

This concept is now extending into the textile area,
specifically carpets, where tufted nylon constructions are
backed with an aqueous polyamide paste laminated with a nylon
backing scrim. At the end of its life the entire carpet can be
depolymerised to caprolactam the raw material for nylon 6.
Similar development is occurring with polyester carpet
constructions.

The aqueous polyamide and ester coatings and adhesives
are based on finely ground polymer usually less than 80µ where
the polymer backbone has been designed to give specific
melting and melt flow properties. These are dispersed in an
aqueous medium with appropriate surfactants and flow control
agents to produce suitable coatings for these products. These
compositions are derived from the garment interlining
industry where they are dot printed via rotary screen printing
techniques onto non woven substrates for subsequent hot
lamination to face fabrics.

1.4 Conclusions

Aqueous coatings for textiles succeed because they
are:
  a) user friendly, both environmentally and in terms of
     compatibility with current textile processing
     equipment and chemicals.
  b) versatile and open to design modification to solve
     problems or to meet new constraints quickly and
     easily.
  c) able to meet most national and international
     specifications.

For textiles the prime requirement is for soft polymers
ie. Tg's <0°C. Most available polymers are tacky under these
conditions and whilst this is advantageous for adhesives this
is not so for coatings. Polyurethanes and silicones for
example can provide many of the required properties but cost
constraints make it difficult to effectively exploit these
products.

Non-aqueous and solvent free are beginning to emerge

but are unlikely to make significant inroads in the foreseeable future until both the chemistry and technology come more into line with the needs of the textile market.

To date, aqueous coating products provide most of the answers required by textile finishers and coaters and look to have an excellent future both in the U.K. and worldwide.

REFERENCES

(1) The Furniture and Furnishings (Fire Safety) Regulations 1988 together with the Amendment 1989.

(2) BS3320 1988 Method for the slippage resistance of yarns in woven fabrics : seam method.

(3) (i) BSEN26330 (1994) Textiles: Determinations of dimensional change to washing and dry cleaning.

   (ii) BS4961. Methods for determination of the dimensional stability of textiles to dry cleaning in tetrachloroethylene. Part 1   1980/1986.

(4) Austrian Standard Ö Norm B6122 Textile glass scrim for outer thermal wall insulation (1988).

(5) German Standard DIN 4102 Part I Fire behaviour of building materials and components.

(6) COSHH: Control of Substances Hazardous to Health Regulations 1988.

(7) CHIPS: The Chemicals Hazard Information and Packaging 1993 (Ref Directive 91/325/EEC and 67/548/EEC.

Waterborne Maintenance Systems for Concrete and Metal Structures

G. A. Howarth

TECHNICAL MANAGER, INDUSTRIAL COPOLYMERS LIMITED, PO BOX 347, PRIMROSE HILL, LONDON ROAD, PRESTON PR1 4LT, LANCASHIRE, UK

1 INTRODUCTION

Concrete and metal, especially steel, are without doubt the two most important construction materials used by man. Protection of structures made from steel and concrete is vital as the cost of deterioration of these materials to the economy runs into billions of pounds. In addition, in recent years the cost to human health and the environment from organic pollutants has also grown immensely.

For many years, two-component epoxy coatings have been used for a wide range of applications including concrete floor sealers and protection of metal from corrosion. After the war until the early 1970s petroleum based products and hence organic solvents were nearly always first choice for coatings and maintenance system. Two-component epoxy products were no exception to this. However, during the 1970s political factors drove up the cost of petroleum based products and around the same time Thomas Swan introduced Casamid 360, a water based polyaminoamide curing agent for liquid epoxy resins which permitted significant reductions in organic solvent content.[1] Since then significant advances have been made in this water based technology and this paper describes these.

2 AMINE AND EPOXY CHEMISTRY

Many amines are now commercially available and these will cure epoxy resins (Figure 1).

The mechanism is by attack of a lone pair of electrons from the amine nitrogen attacking the delta positive carbon with the driving force being the relieving of strain in the oxirane (epoxy) ring.

Use of multifunctional amines and multifunctional epoxy resins will produce a highly crosslinked infusible mass. Examples of multifunctional amines are included in Figure 2:

All these amines are too corrosive and irritant to be used by themselves and are, therefore, reacted with other species to make them less volatile and corrosive.

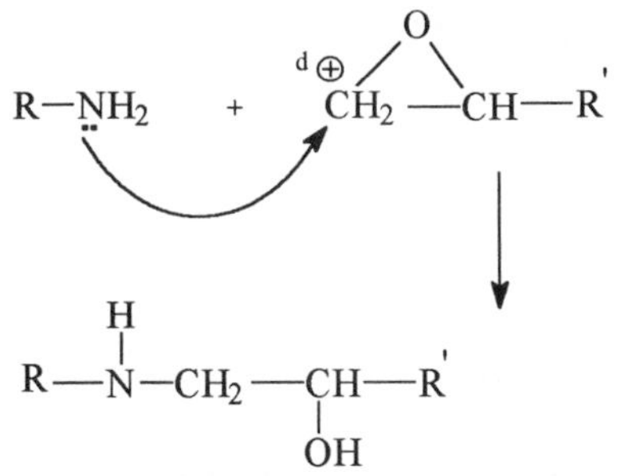

Figure 1 *Amine reacting with an epoxy*

$$H_2N \left(CH_2CH_2NH \right)_X H$$

X=2 DiEthyleneTriAmine   DETA
X=3 TriEthyleneTetraAmine   TETA
X=4 TetraEthylenePentaAmine TEPA

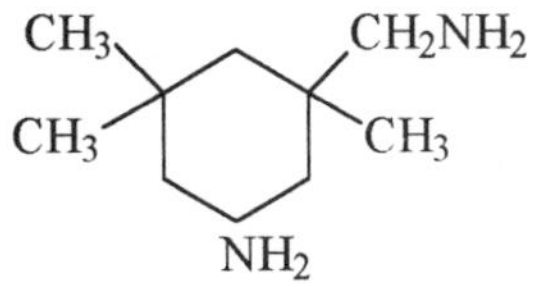

IsoPhorone DiAmine
IPDA

And the polyether amines

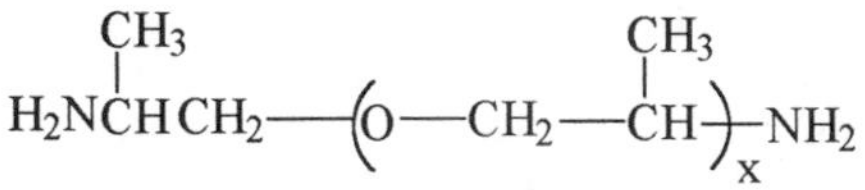

Figure 2 *Some commercially available amines*

2.1 Polyamides

One of the first ways used to convert the amine building blocks into a polymeric species was by reacting with a dicarboxylic acid and condensing off the water (Figure 3).

As polyfunctional amines have been used as the starting materials, these polyamides are usually called polyaminoamides and the remaining amines can be reacted with acetic acid or formaldehyde to give water solubility (Figure 4).

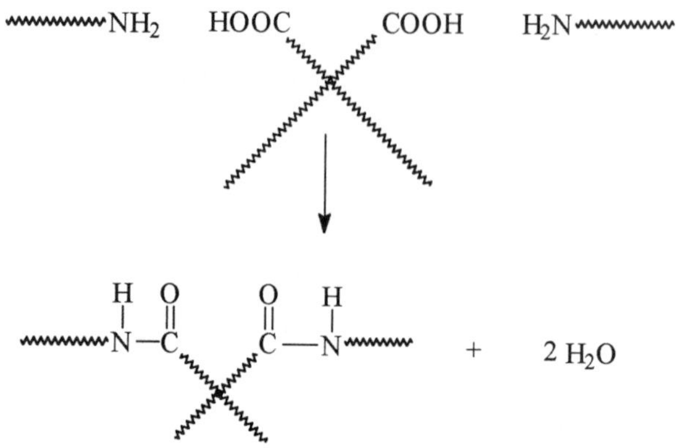

Figure 3 *Amine and acid condensation to give polyamide*

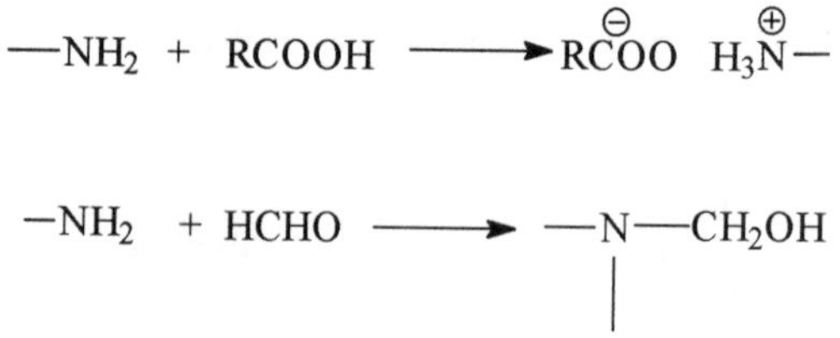

$$-NH_2 \ + \ RCOOH \ \longrightarrow \ RC\overset{\ominus}{O}O \ \ H_3\overset{\oplus}{N}-$$

$$-NH_2 \ + \ HCHO \ \longrightarrow \ -\overset{|}{N}-CH_2OH$$

Figure 4 *Reaction of amine with carboxylic acid or formaldehyde*

A classical soap or surfactant molecule has a long hydrophobic tail and a smaller hydrophilic end. An amine tipped fatty amide is just this and hence the polyaminoamides are inherently surfactant like and will emulsify a liquid epoxy resin. The emulsification process brings the epoxy molecules and amine groups in intimate contact enabling reaction and hence curing and crosslinking to take place. In combination with evaporation of water this produces tough, hard and chemically resistant films.

2.2 Polyamine Adducts

The amine adducts were developed to produce even higher solids and lower viscosity products with a much better colour. The condensation process with the polyamides results in quite a dark colour and if white and pastel shades are required for the metal or concrete coating, then lighter coloured resins are needed. The amine adducts, as the name implies are just amines adducted with other chemical species eg DETA can be reacted with a diepoxide (Figure 5).

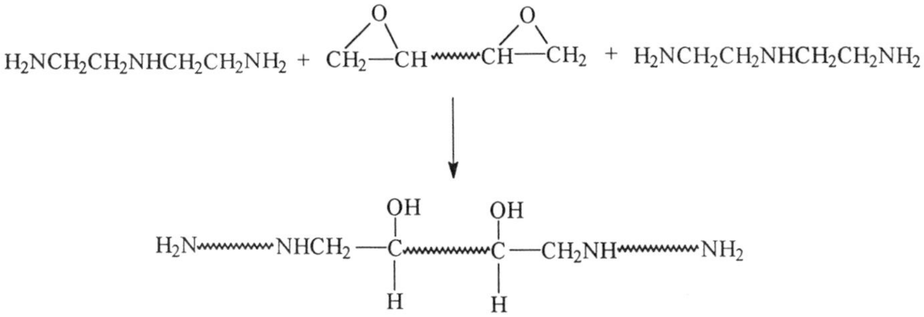

$$H_2NCH_2CH_2NHCH_2CH_2NH_2 + \quad CH_2-CH \text{\scriptsize www} CH-CH_2 \quad + \quad H_2NCH_2CH_2NHCH_2CH_2NH_2$$

$$H_2N\text{\scriptsize www}NHCH_2-\underset{H}{\overset{OH}{C}}\text{\scriptsize www}\underset{H}{\overset{OH}{C}}-CH_2NH\text{\scriptsize www}NH_2$$

Figure 5 *A simple amine-epoxy adduct*

Because there are amine functionalities still present, the molecule can be further adducted with mono, di or trifunctional epoxy species. Water solubility can be enhanced by the incorporation of ether, alcohol, sulphonyl groups, or by the use of glacial organic acid.

It is ideal if glacial organic is not used because over very alkaline substrates such as green concrete, the acid reacts and a very unstable product results especially if the coating is applied by brush. This was certainly one of the weaknesses of the early water based polyaminoamide/epoxy systems. There are a number of polyamine adducts on the market now which are excellent over green concrete. One or two of them have produced an excellent coating over concrete only 7 hours old!

2.3 Polyamine/Polyaminoamide Combinations

The two technologies already discussed can be merged. For example, amine can be partially adducted with an epoxy resin and the chain growth halted by using a monofunctional epoxy (Figure 6).

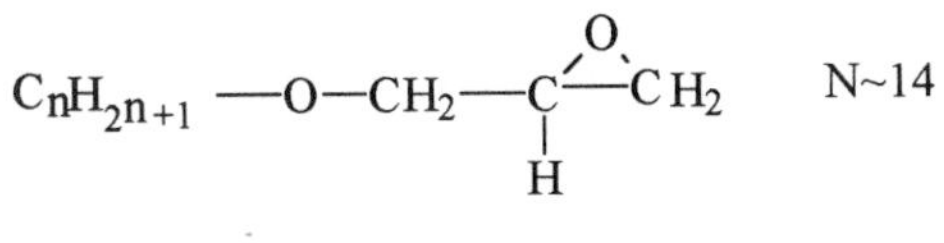

$$C_nH_{2n+1} \quad -O-CH_2-\underset{H}{\overset{}{C}}\overset{O}{\underset{}{\diagup\diagdown}}CH_2 \quad N\sim14$$

Figure 6 *Monofunctional epoxy diluent*

Providing there are a large excess of amine groups, this can be now reacted with a dimer acid and the water produced condensed off to give polyamide characteristics. If it is desired to limit the molecular weight and lower the viscosity, a monofunctional acid can be added to act as a chain stopper. Tertiary butyl benzoic acid is a good example of a chain stopper (Figure 7).

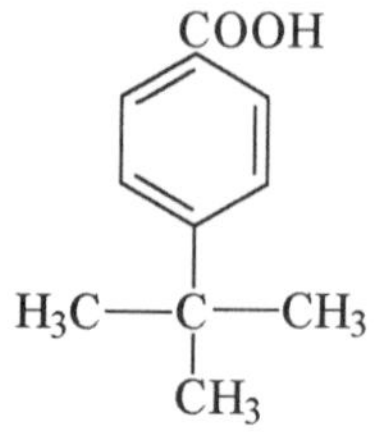

$$\begin{array}{c} COOH \\ \end{array}$$

H_3C—C—CH_3

CH_3

Figure 7 *Para tertiary butyl benzoic acid*

If the product has been formulated so there is still a large excess of amine functionality, glacial acid or formaldehyde can be reacted into the system. As can be seen from this discussion there are an almost infinite number of possibilities for varying the end product properties to individual customer requirements.

The actual chemistry often determines the speed of cure and a waterborne maintenance system for steel or concrete often requires the cure time to be fairly rapid. Quite often a client will want to coat a concrete factory floor on a Friday afternoon and for the floor to be able to take foot and forklift truck traffic on the following Monday morning. This must be borne in mind when formulating.

Another important factor is the level of gloss. Different customers have different requirements. A high level of gloss is still fashionable in the UK but if a factory floor had a very high level of gloss, severe eye strain can result. On the other hand, for waterborne maintenance systems for metals, a very matt primer is often required with a high gloss for the topcoat. The variety of chemistries available in the field of water based epoxy systems enables these differing customer requirements to be met.

2.4 Acrylic Technology

Thousands of papers and many volumes have been written about acrylic technology and the solution and emulsion polymerisation of acrylic monomers. During the late 1980s the technology started to be used in combination with water based epoxy resin technology. Polymers were made with acrylic backbones and tipped with amine. When mixed with epoxy resin the polymer particles coalesced bringing the amine groups in close contact with epoxy groups producing a good film with post crosslinking potential. Carboxylated acrylics have also been produced.[2]

2.5 Newer Epoxy Technology

In addition to the improvements that have been made in curing agent technology, advances have been made with epoxy resins themselves. High molecular weight resins are now available dispersed in water with a small amount of cosolvent. These will dry and film form in their own right by evaporation and coalescence. When mixed with water based curing agent, the amines will post-crosslink to give tough, hard films, which when pigmented yield good maintenance systems for concrete and metal. Very recently a hardener that is pre-emulsified has been developed.

3 EXPLANATION AND DEFINITION OF TERMS

A variety of terms such as Epoxy Group Content, Epoxy Equivalent Weight (EEW), Epoxy Value, Amine Value, Hydrogen Active Equivalent Weight and PHR are used in epoxy resin chemistry.

A summary of the meaning of these terms is worthwhile because they impact on the formulation of waterborne epoxy maintenance systems for concrete and steel.

3.1 Epoxy Equivalent Weight

The EEW is the number of grammes of epoxy resin required to give one mole of epoxy groups (Figure 8).

Epoxy resin Mwt 382

382 grammes gives 2 moles of epoxy groups

EEW =382/2 = 191

Figure 8 *Epoxy equivalent weight*

3.2 Epoxy Group Content

This arises from the EEW and is expressed in the units of mmol/kg ie an epoxy resin with an EEW of 191 has an epoxy group content of:

$$\frac{10^6}{191} = 5236 \text{ mmol/kg}$$

3.3 Epoxy Value

The number of moles of epoxy groups per 100g resin.

eg Epoxy resin Mwt 382

382g gives 2 moles of epoxy resin

$$100g = \frac{100}{382} \times 2 = 0.53$$

Epoxy Value = 0.53

3.4 Amine Value

The number of milligrammes of KOH equivalent to one gramme of hardener resin.

$$\frac{\text{Number of Nitrogens} \times 56.1 \times 1000}{\text{Mwt}}$$

eg TEPA Mwt = 189

$$\frac{5 \times 1000 \times 56.1}{189} = 1484$$

TEPA Amine Value = 1484

3.5 Hydrogen Active Equivalent Weight

When an amine reacts with an epoxy it is the active hydrogen that is involved in the mechanism. A tertiary amine, for example, does not react with epoxy resin though it will catalyse the amine/epoxy reaction. A primary amine can react with two epoxy groups. The definition of the Hydrogen Active Equivalent Weight is, therefore, the quantity of hardener in grammes which contains one mole of active hydrogen.

$$\overset{H}{\underset{H}{\diagdown}}N CH_2CH_2 \overset{\overset{H}{|}}{N} CH_2CH_2 N \overset{H}{\underset{H}{\diagup}}$$

DETA Mwt 103

H Active Equivalent Weight $\dfrac{103}{5} = 20.6$

If DETA were used as a hardener to cure an epoxy resin with an epoxy value of 0.53.

$$0.53 \times 20.6 = 10.92$$

10.92 grammes of DETA would be required.

3.6 PHR

This term is an abbreviation of Per Hundred (of Epoxy) Resin, and is a measure of the stoichiometry of the system. The definition is the weight of curing agent required per 100g of epoxy resin.

For example, using DETA and an epoxy resin of molecular weight 382 from the above calculation the PHR of DETA is 10.92.

For polyaminoamides and polyamine adducts, a range of PHR is usually quoted. Polyaminoamides are very PHR tolerant. Water based epoxy priming systems for concrete are on the market with a PHR of 50 ie with a great excess of epoxy resin. The advantages are that epoxy resins are cheaper than the water based polyamides and hence the system as a whole will be cheaper. In addition it is the epoxy resin which gives the beneficial properties of chemical resistance and toughness and adhesion, therefore, within reason the more epoxy resin, the better. The disadvantage is that the curing time will be longer with excess epoxy.

If a system such as this is used for metal maintenance costings, especially primers, exact stoichiometry is not usually beneficial. Excess epoxy gives good anticorrosion and adhesion and excess amine has been reported to adsorb in the cathodic region and prevent corrosion. However, it is more likely that amines affect both the anodic and cathodic processes.[3,4]

4 EXAMPLES OF WATERBORNE MAINTENANCE SYSTEMS

If concrete itself can be made more chemical resistant, it as the substrate will be the last line of defence in any maintenance system. Up until the 1950s, the idea of adding polymer to concrete has not been considered a commercial proposition as it was considered virtually indestructible. Until very recently, very little research has been done on the synthesis of resins for concrete even though the idea of combining epoxy resin with Portland cement is not new. It is now possible to offer a totally waterborne maintenance system for concrete and metal using water based epoxy technology.[5]

Corroded concrete is removed and the re-bar thoroughly cleaned. A water based epoxy anticorrosive primer can then be applied. The concrete is then made good by applying a water based epoxy concrete repair system. This is followed by a water based epoxy primer for the concrete and finally a water based epoxy topcoat.

4.1 Anticorrosive Primer for Rebar Coating

Table 1 *Guide Formulation   Ref ACP186*

Component	Weight
Incorez 148/024	150.00
Shieldex AC5	50.00
Bayferrox 130M Iron Oxide	200.00
Blanc Fixe N	210.00
Mistron Talc 754	160.00
Tego Foamex 1488	0.50
Nacorr 1651	10.00
Water	419.50
	1200.00
Bisphenol A Epoxy Resin	200.00

There have been a number of articles written about water based anticorrosive primers based on two-pack epoxy technology and the number of systems appearing on the market is increasing all the time.

4.2 Epoxy Modified Mortar Repair System

Table 2 *Guide Formulation Ref EMC-3/148/010*

Component	Weight
Redhill 110	39.15
Pulverised Fuel Ash	14.07
OPC	23.50
Hydrated Lime	1.55
Plasticiser	1.50
	79.77
Incorez 148/010	3.36
Water	13.50
Epikote 828	3.36
	20.22

It can be seen that this formulation is really a 3 pack system. However, newer technology is available whereby the amine adduct and the epoxy resin are adsorbed onto silica and can be added to the mortar mix to make a one pack system that only needs the addition of water. The product can be strengthened further and slumping properties improved by the addition of fibres or other fibrous materials.

4.3 Concrete Primer

Incorez 148/024	300.00g
Epoxy Resin (EEW 191)	350.00g
Water	350.00g
	1000.00g

This primer may be thinned down further with water if required and can be painted on the fresh concrete the following day. This is only a simple guide formulation and the ratio of epoxy resin to hardener may be increased further and tailored precisely to individual customer needs.

4.4 Top Coat

Table 3 *Guide Formulation Ref 95/148 (Modified)*

Component	Weight
Incorez 148/010	360.00
or Incorez 148/024	
Water	317.00
Titanium Dioxide	300.00
Barytes	220.00
Tego Foamex	1.50
Byk 301	1.50
	1200.00
Epoxy Resin	400.00
Total	1600.00

Again this is only a guide formulation with a PVC of 16.3. This may be lowered if desired and the quality improved further by removing the barytes and increasing the resin components accordingly. Also, as with the primer the epoxy to hardener resin ratio may also be increased to improve the chemical resistance further but with a drop off in cure rate.

This repair and maintenance system is thus entirely waterborne and if supplied by the same manufacturer all parts should be completely compatible with each other (Figure 9).[6]

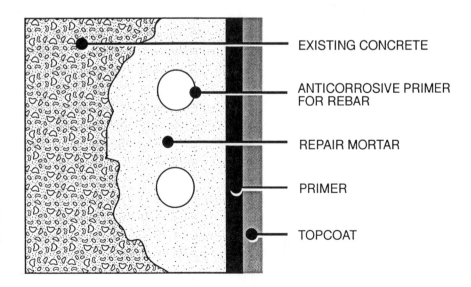

EXISTING CONCRETE

ANTICORROSIVE PRIMER FOR REBAR

REPAIR MORTAR

PRIMER

TOPCOAT

Figure 9 *The Total Waterbased System Approach*

5 NEW ONE PACK TECHNOLOGY

Urethanes have always had outstanding properties such as abrasion resistance. Recently these have become available as dispersions in water. They are single component and, therefore, of interest if they can replace 2 pack waterborne epoxy systems. Very briefly they are manufactured by reacting a multifunctional isocyanate with a polyol and modifying with a diol containing a pendant carboxyl group. This prepolymer is then dispersed into water containing an organic base which reacts with this carboxyl group. The NCO groups are then reacted out by chain extending the product with a polyfunctional amine (figure 10).

Although these products are finding use in the building maintenance industry, there are still one or two areas for improvement.

Polyurethane dispersions are now available entirely cosolvent free, but over highly alkaline substrates such as fresh concrete, the organic base is displaced and instability can results. However, judging by the number of PUD plants being commissioned, their future looks very rosy, though water based two pack epoxies will be around for a long time.

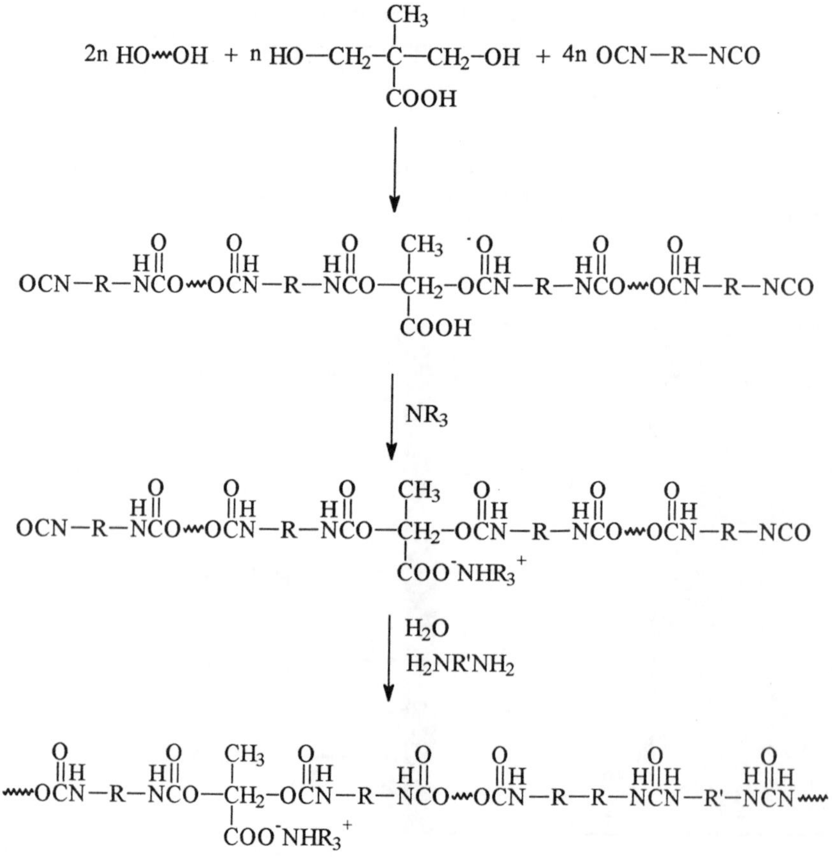

Figure 10 *Polyurethane Dispersion Production*

5 CONCLUSION

Over the years, there have been significant advances in water based epoxy resin and hardener technology. In addition major improvements have been made in other areas of water based technologies. It is now possible to maintain and repair concrete and steel structures with systems that are entirely waterborne. Legislation and customer demands will continue to drive this growth.

References

1 F Richardson, <u>Pigment & Resin Technology</u>, May, 41
2 DOW Chemicals, Data Sheets
3 Editorial, <u>European Coatings Journal</u>, 1994, March, 159
4 Imperial College-London, Lecture Notes
5 L J Daniels, PhD Thesis, University of Lancaster, 1992
6 R F Stanfield, UK Corrosion 93 Paper, Waterborne Maintenance Systems for Concrete Structures from Re-bar to Waterproofing

Crosslinkers

An Overview of Crosslinking in Waterborne Coatings

J. W. Nicholson

MATERIALS TECHNOLOGY GROUP, LABORATORY OF THE GOVERNMENT CHEMIST, TEDDINGTON TW11 0LY, UK

1 INTRODUCTION

Crosslinking of coatings is desirable for a variety of reasons (1). Chemical resistance, resistance to weathering, optimal mechanical properties and barrier properties all improve as molecular weight of the binder goes up. However, applying coatings based on very high molecular weight binders presents severe problems, and for this reason the usual approach is to make paints from relatively low molecular weight polymers, and lightly crosslink them in order to give them the desirable properties in the finished film. Crosslinking should not occur to such an extent that the resulting film is too brittle. Rather the film must retain sufficient flexibility to accommodate movements of the substrate without cracking, flaking or disbonding.

Despite its desirability, crosslinking does not find itself used in all waterborne coating systems. There are other possibilities, as follows :

1.1 Emulsion paints

Colloidal dispersion and emulsion-type paints are very common among waterborne coatings (2), though water-soluble resins are also widely used, for example in automotive applications (3). The term "emulsion" for paints is misleading, since these paints are not based on emulsions at all (4). A true emulsion is a dispersion of one liquid in another (5). Rather these paints should strictly be called "latex" paints, because they are made from a latex, *ie* a dispersion of rubbery polymer in water. However, they are prepared from emulsions of liquid monomer in water, by a process known as emulsion polymerization (6), and this name remains widely used by both consumers and manufacturers.

Drying of emulsion paints involves a complex process of coalescence of the individual latex particles (7). Their effective concentration increases as the water evaporates, which brings the latex particles into contact. They gradually flatten, and merge at the edges to form a film of reasonable continuity. This whole process is complex, and imperfectly understood (8). Nonetheless, it is clear that this insolubilisation process does not involve crosslinking. In principle, if it were possible to remove the coating from the substrate and subject it to sufficiently vigorous agitation, it might be possible to re-disperse the resin and re-establish the original latex. Thus, the drying out of a latex paint is not permanent and may, in principle, be reversed.

1.2 Water-soluble resins

Polymers become soluble in water when they possess a sufficient number of polar functional groups, the more polar the functional groups the better. Poly(acrylic acid), for example, bears a multitude of carboxylic acid groups, and in large enough numbers to make the polymer water-soluble. By contrast, esters of poly(acrylic acid), in which the lightly polar carboxylic acid groups have been replaced by less polar ester groups, are not at all soluble in water. Gradual hydrolysis, though increasing the water solubility, does not confer water solubility until it has progressed to a significant extent. For lightly hydrolysed polymers, the alternative tactic exists of neutralizing the carboxylic acid groups, thereby increasing the polarity of these groups, and increasing their hydrophilic character. As an example, we may consider the copolymer of methyl acrylate and acrylic acid (11% methyl acrylate). It is not naturally soluble in water, but becomes so on neutralization with sodium hydroxide (9). The sodium acrylate units are more polar than the acrylic acid ones, and they are more readily able to overcome the tendency of the hydrophobic methyl acrylate segments to resist dissolution in water.

An important consequence of this relationship between neutralization and water solubility is that reversal of neutralization can be used to confer insolubility as an appropriate waterborne coating resin. Two possibilities exist: (i) electrodeposition, in which an electric current is used to drive the counter ion and the macro-ion apart, causing the macro-ion to interact with water to generate the un-neutralized and hence insoluble form (10); or (ii) volatilization, where the neutralizing species has been ammonia or an amine. In such a case, ammonia or the amine appears in the volatiles, generally following stoving, and the insoluble coating is left behind (2).

In both of these cases, reversal of the neutralization, which is not necessarily permanent, is used to bring about the reduction in hydrophilic character, and thus to convert the coating to an insoluble, water resistant form.

2 CROSSLINKING

Broadly speaking, research in the subject of waterborne coatings concentrates on one of two areas (11):
(i) the development of novel methods of solubilising polymers already known to have film-forming characteristics, and currently used in this way in their native, solvent-soluble form;
and [more rarely],
(ii) the development and exploitation of new crosslinking reactions.

For those paints based on solublised resin systems, crosslinking is typically brought about by the same processes as in conventional paints. These include cure *inter alia* by ring-opening of epoxy rings, and by oxidative drying of conventional drying oils; waterborne coatings have been prepared using each of these processes (2). It is possible to distinguish two groups of crosslinking reactions, those which might be considered general, *ie* reactions that are known for both solvent-based and waterborne versions of particular resins, and those which are specific to waterborne coatings. In the sections that follow, both groups are discussed.

2.1 General Crosslinking

Waterborne coatings are generally cured at elevated temperatures. These range from about 125°C to about 200° depending on the precise details of the resin system and the service needs of the finished paint film. For highly demanding requirements, resins must be well crosslinked, since it is essential that the films lose as much hydrophilic character as possible and become completely insoluble.

Lighter crosslinking can be achieved by the lower temperature heating regimes, usually applied for shorter times. Details of the crosslinking reactions vary with individual resin types, and these are described below:

2.1.1 Drying oils. Paints can be made from conventional solvent-soluble polymers that have been modified to confer either water- solubility or water-reducibility. Drying oils, for example, can be reacted with maleic acid to introduce carboxylic acid groups. These are then neutralized with either amines of ammonia to yield water-soluble oils. Drying of coatings based on such oils is a two step process, in which first the water and the ammonia evaporate, leaving behind an essentially insoluble but tacky film. Full drying is brought about by reaction of oxygen which becomes incorporated into the film from the air. This process is catalysed by metal carboxylate compounds of long-chain fatty acids, in a manner analogous with that of traditional oil paints.

Driers for waterborne oils and their derivatives are similar to those for conventional solvent-based systems, *ie* co-ordination complexes of metals capable of existing in more than one oxidation state. Typically they are based on 1,10 phenanthroline adducts of cobalt, manganese, zirconium and calcium, used either alone or in combination. Lead driers are rarely used, since they pose environmental problems of their own, and do not seem to be any more effective in waterborne systems than the metals mentioned.

2.1.2 Alkyd resins. Alkyd resins have been produced in water-reducible form. Alkyds are formed by esterifying vegetable oils or fatty acids and various mono- and difunctional acids (or anhydrides) with polyols varying in functionality from two to four. The oils and fatty acids can also be of the "drying" type, thus yield air-drying coatings. Alkyds can be made water-reducible in the following ways:

(i) by maleinization of the fatty acid, followed by reaction with dibasic acid and a polyol;

(ii) by making the resin from prepolymers that are rich in hydroxy groups.

Alkyds can be modified to improve the properties of the final paint films, for example by the use of acrylic or silicone prepolymers. In either case, hydroxy or carboxylic acid functional groups are left on the modified resin which both confer affinity for water and act as reactive sites for further chemical reaction en route to fully cured films.

Early drying properties of alkyds tend to be poor, and in order to try to improve them, the following approaches have been made:

(i) To include hard, acidic resins, such as rosin or synthetic polymers having sufficiently high acidic content to confer water-solubility;

(ii) To include small amounts of trimethyol propane triacrylate or other multifunctional acrylate in the alkyd:

$$CH_3-CH_2-C(CH_2O-\overset{\displaystyle O}{\overset{\displaystyle \|}{C}}-CH=CH_2)_3$$

(iii) To copolymerize conventional monomers, such as styrene or methyl methacrylate with the alkyd.

2.1.3 Polyesters. Polyesters can be made and modified by similar techniques to alkyd resins. They differ from alkyds in that they are not made from fatty acids, but from long chain dibasic acids such as adipic acid. The chemistry of their initial prepolymerisation is the same, *ie* condensation leading to the formation of ester links between the monomer units.

The same kind of driers are used in water-soluble alkyds as in waterborne drying oils. In addition, for the alkyds, other driers have been used, including titanium complexes of amino alcohols.

2.1.4 Epoxy resins. Epoxy resins can be made more hydrophilic by the introduction of amine groups using the so-called Mannich reaction (see Figure):

$$R-\text{C}_6\text{H}_4-\text{O-} \quad + \quad CH_2O \quad + \quad H_2N-R'CO_2H \quad \longrightarrow$$

$$R-\text{C}_6\text{H}_4-\text{O-}$$
$$CH_2NHR'CO_2H$$

Figure *Mannich reaction for epoxy resins*

Epoxies of this type can be crosslinked using, for example, amino resins such as methylated melamine formaldehyde. In one such process, cure temperatures of 200-215°C were used, and the final coatings preared as thin lacquers suitable for beer, beverage and food cans (12).

Waterborne epoxies have also been made by incorporating polar functional groups, which have then caused the films to dry without crosslinking. For example, recent work on can coatings has used complex acrylic and methacrylic copolymers with epoxy functionalities. In one case, the resulting copolymer formed stable dispersions, and hence dried by coalescence (13); in the other case, the binder molecules were solublised by neutralization with ammonia and the film dried by volatilization (14).

There are also genuinely crosslinked waterborne epoxies. This may involve the epoxy functionality, as in the case of carbonated polyamido amine curing agents, where gradual reversal of the carbonation process (*ie* loss of carbon dioxide) generates the amine curing agent which then brings about crosslinking through opening of the epoxy ring in the conventional way (15). Most epoxies are hydrophobic, hence require some sort of emulsification if they are to form the basis of waterborne coatings. The use of capped polyamido amines owes part of its success to their effectiveness effect dispersing aids. The crosslinking reaction that is initiated at high temperatures is as follows:

$$R-\underset{\underset{O}{\|}}{C}-O-CH_2CH_2NH-\underset{\underset{O}{\|}}{C}-O-H \quad \longrightarrow \quad R-\underset{\underset{O}{\|}}{C}-O-CH_2CH_2-NH_2 \quad + \quad CO_2$$

carbonated polyamido   polyamido amine –
amine   curing agent for
                              the epoxy resin.

Epoxies have been combined with monomers having conjugated unsaturation, thereby confering "drying oil" character on the resin molecule. In an example of this approach, fatty acids from dehydrated castor oil were heated with maleic anhydride at 200-210°C for six hours under nitrogen (16). They were then further modified by reaction with allyl alcohol to increase the unsaturation and enhance the susceptibility towards oxidative drying. These maleinized fatty acids were esterified with epoxy resins, and the resulting resin used to formulate exterior finished for steel substrates. When dry, these epoxy films were of high gloss, had good anti-corrosive properties, and were able to function as a combined primer and topcoat in a single coat system. The crosslinking of these film followed loss of water, and occured through the oxidative "drying" of the conjugated sites in the drying oil units. Thus this system, though benefiting from the advantages of epoxy resins in terms of general coating performance, actually derived its durability from the drying oil component.

In similar more recent process, coatings were fabricated from copolymers which included epoxy functional monomers together with drying oil moeities and acrylic acid segments (17). The acrylic acid units acted as internal emulsifying sites, enabling the coating formulation to be prepared as an emulsion system.

2.1.5 Polyurethanes. Polyurethanes are polymers formed by reaction of isocyanate groups, -NCO, with hydroxyl groups to yield the so-called urethane group:

$$R-\underset{\substack{\| \\ O}}{C}-\underset{\substack{| \\ H}}{N}-R$$

The isocyanate group itself reacts readily with so-called active hydrogen atoms, which includes those in the water molecule. Consequently, free isocyanate groups cannot be part of any waterborne coating. On the other hand, pre-formed polyurethanes can be used, and here the organic groups can be modified to confer hydrophilic character to the polymer. Modern waterborne urethanes include dispersions comprising one-component, fully reacted polymers that contain only urethane groups, and no free isocyanate. They may be cationic (based on tertiary amines), anionic (based on neutralized carboxylic acid groups) or non-ionic (based on polyethylene oxide) (18). These coatings are used in a variety of applications, including as coil coating primers and in aerospace. These dry by coalscence of the polymer particles, rather than by crosslinking. However, crosslinking of polyurethanes can be introduced by copolymerization with carboxylic acid monomers. In one recent example, a polyurethane-acrylic binder was used in which the free acid groups were reacted with a carbodiimide crosslinker to form the final insoluble film (19).

2.2 Specific Crosslinking

These reactions exploit the presence of those functional groups that confer solubility/dispersibility in water by using them as sites for crosslinking. They are actually very few of these processes in use commercially, though they might reasonably be expected to become more important in the future.

2.2.1 HMMM as crosslinker. External crosslinking agents are frequently used in waterborne coatings, whether applied by electrodeposition or other means, such as spraying. in other words, crosslinking results not from reaction of different functional groups within the resin molecule, but from reaction with a generally low molecular weight additive incorporated into the formulation immediately prior to film application.

A widely used crosslinker is hexamethoxy-methylmelamine, HMMM, which has the following structure:

$$\left(CH_3OCH_2\right)_2N$$

$$\left(CH_3OCH_2\right)_2N \qquad N\left(CH_3OCH_2\right)_2$$

It will react with the carboxylic acid functional groups that solubilise the binder molecules. The resulting crosslinked structure is insoluble, and confers good durability to the finished binder system.

2.2.2 Zirconium crosslinkers: Zirconium compounds have been developed as crosslinking agents for use in paints with carboxylated binders (20). The compounds used include substances such as zirconyl chloride, which tend to form condensed polymeric structures in aqueous solution. For example, at pH 6, zirconyl chloride has the structure:

```
     OH OH   OH OH   OH
      \ | /   \ | /   \ | /
       Zr       Zr       Zr
      / + \   / + \   / + \
         OH       OH
      Cl⁻      Cl⁻      Cl⁻
```

Similar species occur in alkaline solution, *ie* aqueous solutions of compounds such as ammonium zirconium carbonate. These zirconium compounds are able to interact with the functional groups of organic polymers, forming either hydrogen bonds with hydroxyl groups or covalent bonds with carboxylic acid groups.

In one example of the use of a zirconium crosslinker, zirconium acetate was used with a latex comprising butyl acrylate/methyl acrylate/acrylonitrile with 3% acrylic acid as "internal" emulsifying agent. A film was cured by treatment with the crosslinker, followed by heating at 130°C for 30 minutes, after which it showed excellent resistance to acetone. By contrast, a heated film of the latex without any zirconium acetate, showed no resistance to acetone.

An important advantage of using zirconium salts is that they generally have very low toxicity. As a result, ammonium zirconium carbonate is approved by the US Food and Drugs Administration for use in applications involving food contact.

2.2.3 Decarboxylation. In a series of papers, Nicholson and Wilson (21-23) have described in detail the application of the Dakin-West reaction to the cure of polymers containing carboxylic acid functional groups. The Dakin-West reaction is the conversion of carboxylic acids to ketones (24), and is named in honour of the two biological chemists who first studied the reaction extensively as a means of converting α-amino acids to their corresponding ketones (25). To apply this process to the cure of waterborne coatings, a small percentage (*ie* in the range 5-15%) of the carboxylic acid groups are neutralized with a monovalent base, such as NaOH or KOH.

The overall cure process that then occurs in these films is as follows: at elevated temperatures the carboxylic acid groups react to form anhydrides by the simple elimination of water. At the same time, the carboxylate groups undergo loss of carbon dioxide to generate a carbanion site on the polymer backbone. Such reactions are typical of salts of organic acids and are known to be assisted by the presence of nucleophiles. In this system, the carboxylate groups themselves have nucleophilic character, and their presence thus aids the overall cure reaction.

The final step in the process is the reaction of the highly active carbanion sites with the anhydride groups. This produces keto crosslinks and regenerates carboxylate groups. The overall process may be summarised as follows:

$$2 \times R\text{-}CO_2H \longrightarrow \underset{anhydride}{R\text{-}CO.O.OC\text{-}R}$$

and

$$RCO_2^- \longrightarrow \underset{carbanion}{R^-} + CO_2\uparrow$$

then

$$R\text{-}CO.O.OC\text{-}R + R^- \longrightarrow \underset{ketone}{R\text{-}CO\text{-}R} + R\text{-}CO_2^-$$

This process is limited to partially neutralized films of univalent metal polyacrylates. Ammonia did not promote decarboxylation, as it proved too volatile, and what little ammonia remained in the film was converted to amide groups at elevated temperatures. Multivalent ions, such as Co^{2+}, Cu^{2+}, Zn^{2+}, Ca^{2+} or Mg^{2+}, also did not promote decarboxylation. This was shown to arise due to their divalent nature, which inhibits escape of CO_2, rather than to the fact that they form partially covalent bonds and hence bind to particular carboxylate groups in the polymer (26).

Coatings of this type have been prepared from a variety of polymers, including ethylene-maleic acid copolymer (27) and neutralized resins containing predominantly butyl acrylate segments (11). It has been used as an additional crosslinking process in conjunction with esterification, for example with neopentyl glycol and other diols (28). Cure temperatures in the range 180-250°C have been used, mainly in order to develop very high levels of resistance to aqueous media in fairly short stoving times (*ie* of the order of 10 minutes).

   2.2.4 Thermolysis of Bunte Salt polymers.   One method of solubilizing polymers that has been described involves the formation of Bunte salts (29). These are organic thiosulphates, *eg* $R\text{-}S\text{-}SO_3^{-}Na^{+}$, which take their name from Hans Bunte who first described them in 1874. They can be formed from a variety of synthetic procedures, hence can be readily incorporated into polymers.

As originally described, a series of Bunte-type polymers were prepared from poly(epichlorohydrin) modified with amino-ethyl thiosulphuric acid, AETSA (30). The process was not suited to polyvinyl chloride, unfortunately, because this particular polymer undergoes dehydrochrlorination readily under most reaction conditions.   The functional groups of the modified polymers were the free thio-acids, rather than the salts:
ie

$$-NH-CH_2-CH_2-S-\overset{\displaystyle O}{\underset{\displaystyle O}{\overset{\|}{\underset{\|}{S}}}}-OH$$

Depending on the level of incorporation of AETSA, these polymers could be made water-soluble or water-dispersible. They were found to be stable in water for long periods without the need to add external stabilising agents.   These AETSA-modified polymers undergo a complex uncatalysed reaction at 123-135°C involving loss of both SO and SO_2 to yield sulphur bridges:

$$-CH_2-S-S-CH_2-$$

These sulphur bridges act as crosslinks and make the film insoluble in water. Coatings cured in this way have been found to have technically promising properties, though as yet they have not been employed commercially.

3 CONCLUSIONS

An overview has been given of the processes used to crosslink waterborne coatings. Though important in solvent based paints, crosslinking is by no means always used in waterborne systems. Rather, methods of conferring insolubility via a reversal of the water-solubilising process tend to be employed instead, *eg* coalescence of emulsion paints, or regeneration of the less soluble acid or amine functional polymers from their corresponding neutralized forms.

Where crosslinking is used, it is frequently the same as in solvent based paints, *eg* oxidative drying of oils, ring opening of epoxies, and so on. Paints which crosslink by these processes have generally been converted into water soluble or dispersible forms by copolymerization of the parent resin with acrylic or methacrylic acid units, which do not themselves contribute to the crosslinking reactions.

Finally, there are a few crosslinking reactions that are specific to waterborne coatings. These characteristically exploit reactions of those functional groups that confer water solubility. Some, such as reaction of carboxylic groups with HMMM or with zirconium compounds, are of industrial importance, whereas others still await commercial exploitation.

References

1. S. Paul, Ch 6 in "Comprehensive Polymer Science", (Eds G. Allen, J.C. Bevington, *et al*), Volume 6, Pergamon Press, Oxford, (1989).
2. Australian OCCA, "Surface Coatings, Vol 1", Chapman and Hall, London, 1974.
3. Y. Takaya, *Nippon Setchaku Gakkaishi*, 1992, **28**, 296. [Chem. Abs. 1992, **117**, 113547n].
4. W.M. Morgans, "Outlines of Paint Technology, Vol. 2" 2nd edn., Charles Griffen & Co., High Wycombe, 1984.
5. H. Batzer and F. Lohse, "Introduction to Macromolecular Chemistry", 2nd edn., John Wiley & Sons, Chichester, 1979.
6. J.W. Nicholson, "The Chemistry of Polymers", Royal Society of Chemistry, Cambridge, 1991.
7. T. Imoto, *Prog. Org. Coatings*, 1973/4, **2**, 193.
8. J.W. Nicholson, *J. Oil & Col. Chemists' Assoc.*, 1989, **72**, 475.
9. L.J.T. Hughes and D.B. Fordyce, *J. Polym. Sci.*, 1956, **22**, 509.
10. S.R. Finn and C.C. Mell, *J. Oil & Col. Chemists' Assoc.*, 1964, **47**, 219.
11. J.W. Nicholson and A.D. Wilson, *J. Oil & Col. Chemists' Assoc.*, 1987, **70**, 189.
12. C.G. Demmer and N.S. Moss, *J. Oil & Col. Chemists' Assoc.*, 1982, **65**, 249.
13. H. Shimada, H. Takayanagi, K. Nakamura and M. Higuchi, *Japanese Patent* 04,100,871, 2 April 1992.
14. S. Kojima and T. Moriga. *Polymer Science and Engineering*, **33**, 260, (1993).
15 F.B. Richardson, in A.D. Wilson, J.W. Nicholson and H.J. Prosser (Eds), "Waterborne Coatings; Surface Coatings - 3", Elsevier Applied Science Publishers, Barking, 1990.
16. M.M. Shirsalker and M.A. Sivasamban *J. Oil & Col. Chemists' Assoc.*, 1982, **65**, 301.
17. M. Amemoto, *Japanese Patent* 05,117,581, 30 October 1991.
18. R. Arnoldus, in A.D. Wilson, J.W. Nicholson and H.J. Prosser (Eds), "Waterborne Coatings; Surface Coatings - 3", Elsevier Applied Science Publishers, Barking, 1990.
19. G.P. Craun, D.L. Trumbo and F.A. Wickert, *US Patent* 5, 104,928, 14 April 1992.
20. P.J. Moles and J. Byram, *Polym., Paint Col. J.*, 1984, **174**, 440.
21. J.W. Nicholson and A.D. Wilson, *Br. Polym. J.*, 1987, **19**, 67.
22. J.W. Nicholson, R.P. Scott and A.D. Wilson, *J. Oil & Col. Chemists' Assoc.*, 1987, **70**, 159.
23. J.W. Nicholson and A.D. Wilson in A.D. Wilson, J.W. Nicholson and H.J. Prosser (Eds), "Waterborne Coatings; Surface Coatings - 3", Elsevier Applied Science Publishers, Barking, 1990
24. G.L. Buchanan, *Chem. Soc. Revs.*, 1988, **17**, 91.
25. H.D. Dakin and R. West, *J. Biol. Chem.*, 1928, **78**, 91.
26. J.W. Nicholson, E.A. Wasson and A.D. Wilson, *Br. Polym. J.*, 1988, **20**, 97.
27. J.W. Nicholson and E.A. Wasson, *Br. Polym. J.*, 1989, **21**, 513.
28. J.W. Nicholson and E.A. Wasson, *Polym. Paint. Col. J.*, 1989, **179**, 397.
29. S.F. Thames, in A.D. Wilson, J.W. Nicholson and H.J. Prosser (Eds), "Waterborne Coatings; Surface Coatings - 3", Elsevier Applied Science Publishers, Barking, 1990.
30. S.F. Thames, *J. Coatings Tech.*, 1983, **55**, 87.

Development and Application of Waterborne Radiation Curable Coatings

W. D. Davies and I. Hutchinson

AKCROS CHEMICALS, PO BOX 1, ECCLES, MANCHESTER M30 0BH, UK

1 INTRODUCTION

Radiation curing systems made a big impact in the late 1970's as a solution to both energy conservation and environmental pollution in the ink and coating industries. Since the original systems were essentially 100% solids it was not initially perceived as advantageous to develop water based systems. In recent years, however, there has been a steady growth in the development and use of these systems.

This paper will discuss some of the reasons for this growth as well as reviewing some of the more important product types and methods of manufacture.

Particular attention will be focused on polyurethane based systems. Some of the key application areas for this class of products will be considered along with the details of formulation and performance testing.

2 WHY WATERBORNE RADIATION CURABLE SYSTEMS?

Radiation curable systems were first commercialised in the early 1970's and they have enjoyed very good growth during the last two decades.

Current estimates for the use of these systems are:-

European $\simeq$ 27,000 t.p.a.
Worldwide $\simeq$ 65,000 t.p.a.

The success of radiation curable systems can be attributed to a number of factors of which the following are amongst the most important:
- Low energy requirement for curing
- Low VOC - most systems are 100% solids
- Fast curing
- Suitable for use on heat sensitive substrates
- Space saving

The reasons for the acceptance of water based radiation curable systems is now very well appreciated though initially it seemed a retrograde step. To fully appreciate the role of water in these systems, one needs to look at a typical U.V. curable formulation.

	p.b.w.
Resin	70
Reactive diluent	25
Photoinitiator	3
Other additives, e.g. flow aids, matting agents etc.	2

There are a legion of different types of resins that can be used in radiation curable systems. The most popular types are acrylate tipped epoxies, polyester or polyurethanes. Acrylated epoxies are derived from the reaction of an epoxy resin with acrylic acid whilst polyester acrylates are made by the standard condensation reaction between polyols and mixtures of acrylic and poly acids. Acrylated polyurethanes are discussed in Section 2.3.

Reactive diluents or monomers as they are sometimes referred to, are normally simple acrylate esters such as tripropyleneglycol diacrylate or trimethylol propane triacrylate.

The amount of reactive diluent required is often dictated by its ability to reduce the viscosity of the resin to that demanded by the application method for a particular process. In many applications such as wood finishing a low viscosity is required so that the amount of material coated onto the wood is low enough not to deter from the aesthetic appeal of the product. Clearly there are limits on the amount of reactive diluents that can be used in a given formulation which will be determined by considering the effect of the reactive diluent on the physical properties of the final cured coating as well as with health and safety aspects. Low molecular weight acrylates of the type used as reactive diluents are potential skin irritants though they do vary considerably in the degree of severity. Considerable strides have however, been made by manufacturers of these reactive diluents to produce less irritating grades.

A further problem with a high monomer content system is that in some applications the substrate can preferentially absorb the monomer during application. Since U.V. radiation has limited penetration power this results in undercuring of the system which can give rise to poor physical properties, such as adhesion. In addition problems of odour may arise from the uncured monomer.

The nett result however is that there is a requirement to replace (either fully or in part) the amount of reactive diluent and it is here that water can fulfil an important role. Water can be used either as the diluent or the dispersing medium for a given resin thus we have two main types of water based systems.

1. Water reducible
2. Water based dispersions

This paper will concentrate on the latter type, however, the water reducible types will be discussed briefly in the next section.

2.1 Water reducible systems

In this type of system the resins are modified chemically so as to be able to accept a certain amount of water which will act to reduce the viscosity. The rheological properties of this type of system resemble that of a solution and therefore the viscosity will depend very much on the molecular weight and concentration of the resin. In general the introduction of the high concentration of hydrophilic groups necessary to achieve water solubility leads to a reduction in the physical properties of the final cured resin.

A balance has to be struck when modifying the resin so that it does not become so hydrophilic that even the final cured coatings are also susceptible to attack by water.

Bate and Salim (1) are among many workers who have reported their work on the application of water reducible poyurethane systems.

2.2 Water based dispersions

Aqueous dispersions of acrylate tipped resins have similar characteristics to any other type of aqueous dispersions from the rheological point of view, they display the usual features in that the viscosity is to a large extent independent of molecular weight of the resin.

Dispersions fall into different categories depending on the 'charge type' and also the means employed in their manufacture. The two main techniques for manufacturing dispersions are:

1. external emulsification
2. internal or self emulsifiable system

In the first case a given resin is emulsified in water using one or more emulsifiers, the choice of which depends very much on the resin type. The issues to be addressed in choosing a surfactant are also very much tied up with the actual mechanical aspects of emulsification. The many factors that have to be considered are of course not specific to acrylate tipped resins hence the general rules for emulsification of resins apply.

Numerous articles have been published on the subject of "resin emulsification" (2,3,4).

It should also be noted that there exists the possibility of using hybrid systems in which a resin with no acrylate functionality is co-emulsified with either another acrylate tipped resin or a multifunctional acrylic monomer.

2.3 Self emulsifiable dispersions

In this class the resin in question is suitably modified by building in groups than will render it emulsifiable in water. There are several ways open to the chemist to modify the resin so as to render it self emulsifiable and these methods have been reviewed in other publications (5,6,7,8). Undoubtedly the most exploited has been the use of carboxyl groups which can be neutralised and subsequently emulsified in water. The rapidly growing family of polyurethane dispersions now being offered are often made using this principle. Indeed, the method of manufacturing U.V. curable urethane acrylates is merely an extension of the well established procedures for preparing the thermoplastic types which have been commercialised for well over 20 years.

The high viscosities and low solubility of P.U. acrylates in reactive diluents means that as a class they benefit more than any other from being produced in the form of an aqueous dispersion.

An outline process for the manufacture of an anionic urethane acrylate dispersion will now be presented.

WATER BASED URETHANE ACRYLATE DISPERSIONS
METHOD OF MANUFACTURE

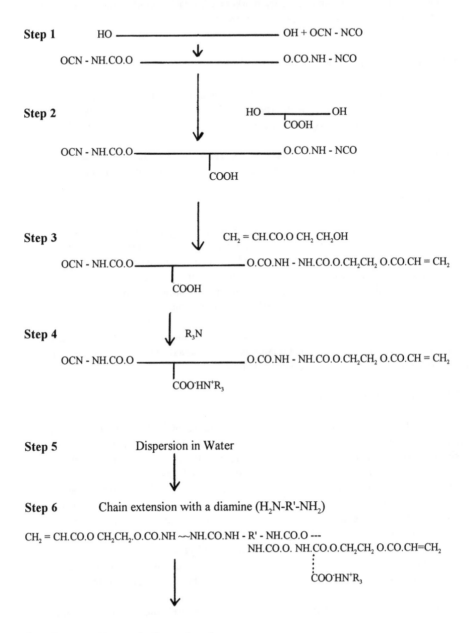

Step 1   HO ——————————————— OH + OCN - NCO

OCN - NH.CO.O ——————— ↓ ——————— O.CO.NH - NCO

Step 2   HO ————— OH
                                                              |
                                                            COOH

OCN - NH.CO.O——————————————— O.CO.NH - NCO
                                  |
                                COOH

Step 3   ↓   CH$_2$ = CH.CO.O CH$_2$ CH$_2$OH

OCN - NH.CO.O——————————————— O.CO.NH - NH.CO.O.CH$_2$CH$_2$ O.CO.CH = CH$_2$
                                  |
                                COOH

Step 4   ↓   R$_3$N

OCN - NH.CO.O ——————————— O.CO.NH - NH.CO.O.CH$_2$CH$_2$ O.CO.CH = CH$_2$
                                  |
                           COO⁻HN⁺R$_3$

Step 5   Dispersion in Water
                                  ↓

Step 6   Chain extension with a diamine (H$_2$N-R'-NH$_2$)

CH$_2$ = CH.CO.O CH$_2$CH$_2$.O.CO.NH ⁓NH.CO.NH - R' - NH.CO.O ---
                                     NH.CO.O. NH.CO.O.CH$_2$CH$_2$ O.CO.CH=CH$_2$
                                          ⋮
                                     COO⁻HN⁺R$_3$

                                  ↓

Step 7   Removal of organic solvents

Step 1. This stage involves the preparation of a conventional polyether or polyester based isocyanate terminated prepolymer. Typically a difunctional polyol and a diisocyanate is used.

Step 2. At this stage the carboxyl function is introduced. Dimethylol propionic acid is one of the most widely used sources of carboxyl groups. Organic solvents may also be introduced at this stage to facilitate handling of the prepolymer.

Step 3. The acrylic function is incorporated by reaction of 2-hydroxy-ethyl acrylate with the isocyanate terminated prepolymer. The degree of end capping can be varied, leaving some free NCO groups for further reaction.

Step 4. The prepolymer is neutralised with a tertiary amine such as triethylamine.

Step 5. Neutralised prepolymer is dispersed in water.

Step 6. Residual NCO groups from Step 3 are chain extended with a diamine such as ethylene diamine.

Step 7. Removal of any organic solvents introduced earlier in Step 2.

The vast range of polyols and isocyanates available as building blocks coupled with many variables that can be brought about with the stoichiometry and process conditions makes polyurethanes one of the most exciting class of products to work with.

The processes for manufacturing dispersions and in particular the polyurethane types is often very sensitive to both the processing conditions employed as well as the composition. The effort in preparing dispersions is however readily justified when examining the advantages that these systems offer to the end user in terms of improved processing and performance. Some of these advantages are recorded below:

Advantages of waterborne systems
- Greater flexibility in design of base resins
- Easier to formulate e.g. incorporation of matting agents
- Can be applied by spraying
- Coating is dry at the pre- U.V. curing stage
- Skin irritancy/sensitization problems associated with reactive diluents are eliminated
- Low coating weights can be easily applied - important in inks and wood finishing
- Easy to clean off application equipment
- Low odour
- Low shrinkage

3 PHYSICAL PROPERTIES AND APPLICATION

3.1 Physical properties

The properties that are usually measured to characterise dispersions are: solids content, pH, viscosity and particle size.

Typical values for a polyurethane acrylate dispersion are given in Table 1.

Table 1 *Typical Physical Properties of Aqueous Urethane Acrylate Dispersions*

Appearance	Translucent/milky white
% Solids	40
pH	8.0
Particle size	0.1 micron
Viscosity (25°C)	15 seconds (Ford cup No. 4)

A wide variety of parameters can be measured on the derived cured coating and these vary tremendously depending on the type of base resin used in the dispersion. The choice of resin is very much dependent on the particular end application for the coated article. Invariably, the cost of the dispersion and the performance characteristics of the coating go hand in hand.

In this paper dispersions of urethane acrylates have been discussed at some length since it is generally acceptable that this class does offer the best physical properties. It is also true to say that they tend to be the most expensive and therefore will only be used in applications that demand high performance and thus accommodate the price premium. The flexibility in terms of design of resin means that the scope for using urethane acrylates is much wider than for any other class.

It is traditional in the field of conventional urethane coatings, e.g. moisture curing types, to quote the physical properties of unsupported films. However, for U.V. cured coatings it is considered that for most applications, it is more relevant to measure the properties of the coated article.

One very practical reason for doing this is that it is very difficult to measure accurately the properties of thin films such as those encountered in many applications (25 microns or less).

The following is a list of some of the main application areas for waterborne U.V. curable systems plus the key physical properties associated with them.

Application area	**Key parameter**
Screen inks	Rub resistance
Wood finishing (furniture)	Gloss, solvent resistance, sanding
Wood finishing (floors)	Gloss, solvent/stain resistance, abrasion
PVC floor coating	Solvent/stain resistance, scuff, light stability
Textiles	Flexibility/soft handle
Leather	Flexibility

3.2 Formulating water based U.V. curable systems

In this section the results of experimental work carried out on various waterborne resins will be presented and discussed. Various methods of assessing the performance of coatings will also be illustrated.

3.2.1 *Sources of radiation.* The two main sources of radiation used in the ink and coatings industries are electron beam and ultra violet (U.V). The latter is more important commercially and the primary reason for this being the prohibitive cost of the EB equipment for most applications.

In this paper results will be presented for both U.V. and EB cured coatings. The specific equipment used was as follows:-

EB   A laboratory electrocurtain unit manufactured by Energy Science.

U.V.   1. Conventional mercury vapour arc lamp, output 120 watts/cm manufactured by Wallace Knight Limited.
       2. Electrodeless mercury vapour lamps, output 120 watts/cm manufactured by Fusion Systems.

3.2.2 *Photoinitiators.* The photoiniator is a key ingredient in any U.V. curable formulation and there are many types available commercially. There are several publications that discuss the relative merits of the various types of photoinitiators, including water soluble types (9,10,11,12).

A photoinitiator may well be very specific in its performance characteristics depending on the nature of the resin and the curing conditions employed. There is no substitute to carrying out a screening exercise to identify the 'best' photoinitiator for a given application. Some of the issues that have to be considered are:

· Solubility in the resin
· Optimum concentration
· Matching absorbance characteristics with source of radiation
· Pre-drying conditions
· Temperature

3.2.3 *Test methods.* In order to assess the performance characteristics of a coating system a wide spectrum of tests may be used. These vary from ones that are specific to a particular end application on the one hand to ones that are totally independent on the other. Examples of some tests are given below:

Table 2 *Performance Testing*

Spectroscopic techniques	Infra red measurements
Physical properties of cured films	Tensile Strength Thermal analysis (DMA)
Chemical Resistance	Resistance to solvents, acids, bases, various stains
End use related tests	Scuff, adhesion, flexibility, hardness, abrasion

The practical aspects of some of the test methods referred to in Table 2 are discussed briefly in the following section.

3.2.3.1 *Infra red spectroscopy (IR)*. The classical approach to using IR as a means of assessing the degree of cure of U.V. cured systems is to measure the reduction in the absorption of the 810 cm^{-1} band associated with C-H out of plane deformation of the acrylic double bond of coatings cured under a variety of conditions. Attenuated Total Reflectance (ATR) is a convenient way of evaluating coated substrates, but it must be realised that this technique only measures the surface of the coating. Since it is known that the degree of cure can vary with film thickness, it is important, therefore to eliminate this variable by comparing the absorbance of the 810 peak with that of a suitable internal reference peak which is not associated with the acrylate group.

The results obtained from ATR measurement must be interpreted with care as they do not give any indication about the effectiveness of the through cure of a given coating. This is particularly important when comparing photoiniators since some types are better for surface cure whilst others are better for through cure.

In order to illustrate the use of ATR the performance of 5 commercially available photoinitiators was compared in a standard U.V. cured formulation using 1% b.w. on a 40% solids P.U. acrylate dispersion. The actual photoinitiators employed are listed in Table 3.

Table 3 *Photoinitiators*

Ref No.	Trade Name	Chemical Name
A	Esacure KIP100F	30/70 mixture of 2-hydroxy-2-methyl-1-phenyl propane-1-one and oligo [2 hydroxy - 2 - methyl -1 4-(1 methylvinyl) phenyl] propanone
B	Darocur 4265	1/1 mixture 2-hydroxy - 2-methyl - 1 - phenyl propane - 1 - one and 2,4,6 - trimethyl benzol diphenyl phosphine oxide
C	Irgacure 500	1/1 mixture of 1-hydroxy - cyclohexyl - phenyl - ketone and benzophenone
D	Irgacure 2959	1-[4-(2-hydroxyethoxy) - phenyl] - 2-hydroxy - 2 - methyl - 1-propane-1-one
E	Darocur 1173	2-hydroxy - 2 methyl-1-phenyl propane-1-one

Each formulation was applied to PVC, cured and the ATR spectrum for the acrylate double bond plus the internal reference peak measured. The actual IR absorbance is expressed as a ratio of the two absorbant peaks. The higher the value of the ratio the higher the double bond content. Examining the results in Table 4 it can be seen that photoinitiator A has given the best results, i.e. highest conversion of double bonds. This photoinitiator was then examined in more detail and the effect of concentration and intensity of U.V. radiation was studied and the results of this investigation are presented in Figure 1.

FIGURE 1

IR ABSORBANCE VS PHOTOINITIATOR CONCN./U.V. DOSE

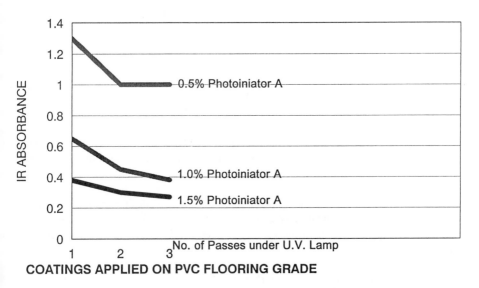

COATINGS APPLIED ON PVC FLOORING GRADE

FIGURE 2.

INFRA RED ABSORBANCE VS. E.B. DOSAGE

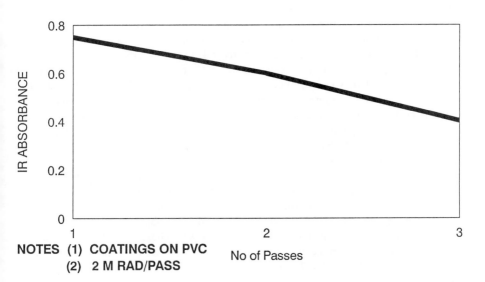

NOTES (1) COATINGS ON PVC
(2) 2 M RAD/PASS

Table 4 *Infra Red Absorbance vs Photoinitiator Type*

Photoinitiator	Infra Red Ratio
A	0.3
B	0.48
C	0.57
D	0.32
E	0.65

Notes: (1) **Coatings on PVC**
               (2) **Initiator 1% w/w on dispersion**

A similar exercise was carried out to study the effect of E.B. radiation, the main difference being that the photoinitiator was omitted from the formulation. The results are presented in Figure 2.

3.2.3.2 *Film hardness.* This is one of the simplest techniques for measuring cure and is best carried out by applying the coating to glass or metal. A pendulum tester, such as the Konig is then used to measure the hardness of coatings cured under different conditions.

3.2.3.3. *Dynamic mechanical analysis (DMA).* DMA is one of the most powerful techniques for measuring the glass transition temperature (Tg) of a polymer. In the work reported in this paper glass braid was dipped into a water based U.V. curable system based on a P.U. acrylate dispersion and the water dried off followed by curing with U.V. The modulus temperature profile of the polymer is then obtained using a Thermal Analysis DMA analyser Model 983 (T.A. Instruments) from which the Tg is obtained (Figure 3).

Results will now be presented demonstrating the use of DMA to assess the effect of processing conditions on cure. Let us first of all consider the water removal or drying stage. It is this stage that is the main difference between waterborne and conventional U.V. cured systems and has been the subject of much discussion in numerous publications (8,9,10). Clearly from a processing point of view it is an obvious drawback of having to install some form of heating, adding both to the installation and subsequent running costs. There is also another important issue to be considered namely that of the potential loss of the photoinitiator during the drying stage. Some workers have observed an adverse effect of drying at elevated temperature which is dependent on the photoinitiator type and the curing conditions (10,13).

Initially, work was carried out on formulations based on photoinitiator A. The coated glass braid was dried at relatively mild conditions of 65°C for 15 minutes and the Tg of the system was then measured before and after U.V. curing. This exercise was then repeated but this time the coated glass braid was subjected to an extra thermal process of 190°C for 2 minutes. These conditions were chosen as it has a relevance in the coating of PVC when the normal processing conditions for the expansion or blowing stage is typically 190°C for 2 minutes. The results for this exercise is recorded in Table 5.

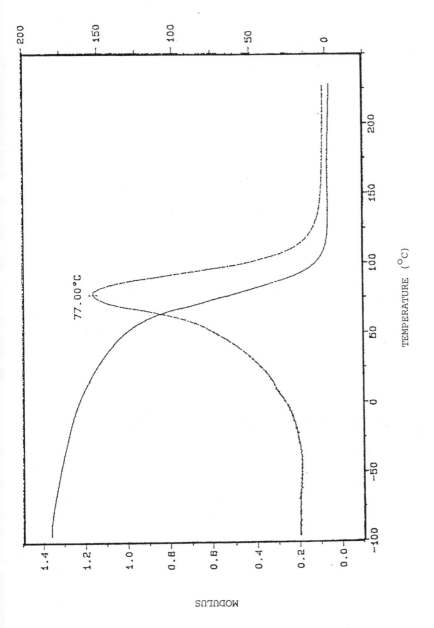

FIGURE 3   DMA MODULUS TEMPERATURE PROFILE

Table 5 *Tg Values (DMA) at Different Drying U.V. Curing Conditions*

U.V. Curing Conditions		Pre U.V. Drying Conditions			
		65°C/10 mins		190°C/2 mins	
Lamp	**U.V. Dose (milliJoules)**	**P.I. (A) Tg(°C)**	**P.I. (E) Tg(°C)**	**P.I. (A) Tg(°C)**	**P.I. (E) Tg(°C)**
None	-	32	-	46	44
Fusion	280	74	-	77	49
Fusion	790	76	-	104	-
Wallace Knight	790	76	-	105	79
Wallace Knight	790 + IR	77	-	101	97

The first point to note is that the Tg of the uncured resin increases from 32°C to 46°C as a result of the extra thermal treatment at 190°C. The most likely reason for this is that the residual water in the first case acts as a plasticizer thus reducing the Tg. It is well known that removal of the last traces of water from anionic dispersions is difficult.

The effect of the drying conditions continues to manifest itself when examing the U.V. curing stage. For instance the coating based on photoinitiator A dried at the lowest temperature yields Tg values of 74°C at a U.V. dose of 280 m joules and it only increases to 76°C at a dose of 790 milliJoules, whereas corresponding Tg's for coating subjected to 190°C/2 mins are 77°C and 105°C respectively.

It is interesting to note that for photoinitiator E which gave the worst results from the ATR studies, the use of IR heaters prior to the U.V. exposure results in a system with a Tg of 97°C which is very similar to the corresponding system based on photoinitiator A having been subjected to the same drying and curing conditions.

Based on these observations it can therefore be concluded that the use of high temperatures during the drying cycle has a beneficial rather than any detrimental effect.

3.2.3.4 Solvent and chemical resistance. Many applications, e.g. wooden or PVC floors, demand a significant degree of resistance to attack by certain chemicals. Results of a selection of such tests are presented in Table 6 and 7 for both U.V. and E.B. cured coatings applied to primed beechwood. These results demonstrate urethane based products have very good resistance overall. It is also interesting to note that when the same coating formulation was applied to steel, the MEK rub resistance tests were very poor. The results for coatings on steel are given in brackets. This effect can be attributed to the poor adhesion of the coating to steel. When carrying out chemical resistance tests such as the ones described in this section on laboratory produced test pieces, the coating may have imperfections - pin holes etc. which can result in misleading results.

Table 6 *Performance Testing of Coatings on Wood (U.V.)*

Ref	P.I.	No. of Passes	Cross Hatch (%) Adhesion	MEK Double Rubs	Acetone	10% Acetic Acid	20% Sodium Hydroxide	Black Shoe Polish
2923/50C/1	A	1	100	>100(5)	Slight Damage	No change	No change	No stain
2923/50C/2	A	2	100	>100(31)	Slight Damage	No change	No change	No stain
2923/50C/3	A	3	100	>100(20)	Faint Damage	No change	No change	No stain

Table 7 *Performance Testing of Coatings on Wood (E.B.)*

Ref	No. of Passes	Cross Hatch (%) Adhesion	MEK Double Rubs	Acetone	10% Acetic Acid	20% Sodium Hydroxide	Black Shoe Polish
2923/52/1	1	100	45(<45)	No change	No change	No change	No stain
2923/52/2	2	100	44(<5)	No change	No change	No change	No stain
2923/52/3	3	100	55(<5)	No change	No change	No change	No stain

Interpretation of results

Not surprisingly, when comparing results from a variety of methods the conclusions that can be drawn are not always the same. Each set of results must therefore be interpreted carefully as to what is being measured and its relevance to the end application. Cure and conversion for instance are two terms that are often confused.

Cure has been expressed in many ways but ultimately it should be viewed in similar way to 'quality', which has been defined as "fit for purpose".

4 CONCLUSIONS

Initially industry was reluctant to consider the use of water based radiation curing systems but as benefits such as ease of processing coupled with the technical merits became more apparent, this class of products is now well established.

The flexibility of polymer design associated with the use of dispersions, together with the eloquence of UV or EB radiation as the means of crosslinking will ensure continued growth in existing applications as well as penetration into new markets.

5   ACKNOWLEDGEMENTS

The authors wish to acknowledge the valuable comments and suggestions received from many colleagues during the preparation of this paper.   The contribution of the following people in particular is appreciated.

Ms. Claire Meredith and Mr. John Hynds -		Coating and formulating
Mrs. Sylvia Massey	-	Infra red spectroscopy
Dr. Jim Mahone	-	Thermal analysis

References

1. M.S. Salim and N. Bate, *Polymer Paint Colour J.* 1989, **179** (4239) 400.
2. G. Dahms and O. Hafner, *Paint and Resin* 1988, August, 13.
3. P.A. Reitano and T. Szurgyjlo, *Modern Paint and Coatings*, 1990, July, 44.
4. A. Bouvy and G.R. John, *Polymer Paint Colour J.* 1993, **183** (4330) 284.
5. M. Phillips et al, *Polymer Paint Colour J.* 1993, **183** (4322) 38.
6. R. Arnoldus, *Polymer Paint Colour J.* 1990, **180** (4263) 434.
7. F.J. Kosnik, *Presented at Water-borne and High Solids Coating Symposium*, New Orleans, February 1989.
8. P.H. Stenson, *Radtech Report.* 1993, January/February, 18.
9. G.L. Bassi and F. Broggi, *Polymer Paint Colour J.* 1988, **178** (4210) 197.
10. W.A. Green, *Polymer Paint Colour J.* 1994 **184** (4358) 474.
11. C. Armstrong, *Eurocoat J.* 1994, **4**, 178.
12. D. Skinner, *Polymer Paint Colour J.* 1994, **184** (4361) 566.
13. M.J. Dvorchak and B.H. Riberi, *Presented at the Federation of Societies for Coatings Technology* 1990 in Washington D.C.

Chemistry of Amino Resins and Their Cross-linking Mechanisms

R. McD. Barrett

BIP SPECIALITY RESINS LIMITED, PO BOX 3158, ROOD END ROAD, OLDBURY, WARLEY, WEST MIDLANDS B69 4HL, UK

INTRODUCTION

Amino resins were commercialised in the 1930's. The original products were simple condensation products of urea and formaldehyde and were used as adhesives for wood and as crease resistant finishes for textiles. In their early applications the resins were not, strictly speaking, cross-linking agents, but were used as bulk polymers. Melamine formaldehyde resins were available in the 1940's and their introduction further expanded the applications for amino resins.

At first sight it is surprising that products which were developed so long ago are still of sufficient importance to merit discussion today. Nevertheless they remain a very important class of industrial chemical, as witnessed by the annual production, which is in excess of 1×10^6 tonnes world-wide. The longevity of amino resins is a consequence of their relative low cost and their versatility, which has enabled them to keep pace with current technical demands.

It is also surprising that, despite the age of melamine resins, the reaction mechanisms involved in reaction with other polymers are not fully understood. A recent paper suggested some of the reasons for the lack of understanding were:

1.   Melamine resins and their co-reactant polyols are complex mixtures.

2.   Each component of these mixtures can undergo a variety of reactions.

3.   The presence of water at varying levels influences the reactions.

4.   Most, and possibly all of the crosslinking reactions are reversible.

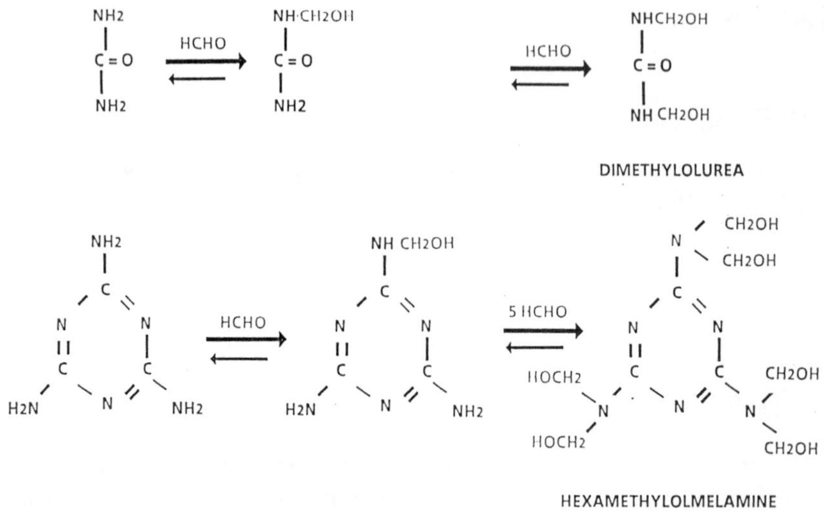

Figure 1   *Addition of formaldehyde to form methylol groups*

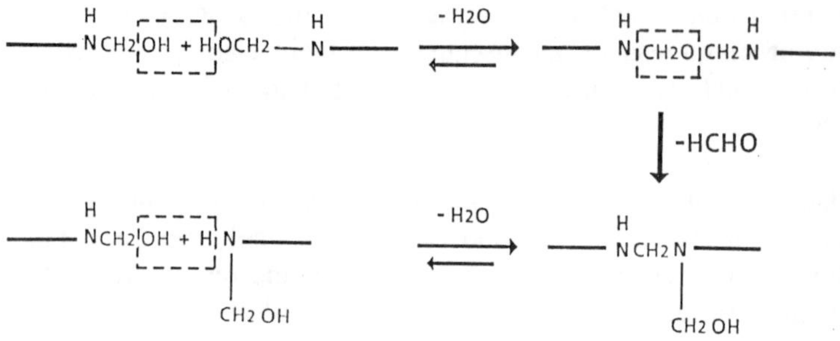

Figure 2   *Condensation reactions*

PREPARATION

The first stage in the synthesis of amino resins is the addition of formaldehyde to amide groups, to form methylol compounds (see FIGURE 1). This reaction takes place at neutral or alkaline pH, even at room temperature. Since the reactions are reversible, although the products are solids, it is difficult to isolate pure products. Using melamine, it is possible to produce a range of methylol derivatives from the momo to hexa. Hexamethylol melamine is insoluble in water in consequence of hydrogen bonding.

Condensation polymerisation of methylol groups occurs under acidic conditions with the progressive build up of molecular weight, and without the loss of functionality (see FIGURE 2).

Complete reaction of all functionality would lead to the formation of a cross-linked gel as shown in FIGURE3.

Although in practice reaction does not go to completion, nevertheless a hard, clear, brittle and relatively inert solid is formed.

The condensation reaction occurs stepwise and can be stopped by neutralisation when the required molecular weight and viscosity have been obtained. However, polymethylol urea and melamine resins are not very stable and their use is limited. It is possible to modify the properties by etherifying methylol groups as shown in FIGURE 4. The conditions which promote the etherification reaction - heat and acidity - also produce condensation, so that both reactions will occur simultaneously. Varying the concentration of alcohol can be used to control the degree of etherification. These etherified amino resins are a very important class of crosslinking resins. Compared to methylol amino resins, the etherified resins generally have better stability, solubility and compatibility with other polymers. Depending on the etherifying alcohol, either water soluble or solvent soluble polymers can be produced. Detailed resin compositions will be considered later.

CROSSLINKING

The primary application of etherified amino resins is in the surface coatings industry where they are used as crosslinking agents in combination with hydroxy functional polymers (see FIGURE 5). The cross-linking reaction is facilitated by the application of heat after the paint has been applied. Typical stoving schedules range from 30 minutes at $120°C$ to 10 seconds at $230°C$. Common end uses are automotive paints, can coatings for food and beverages,

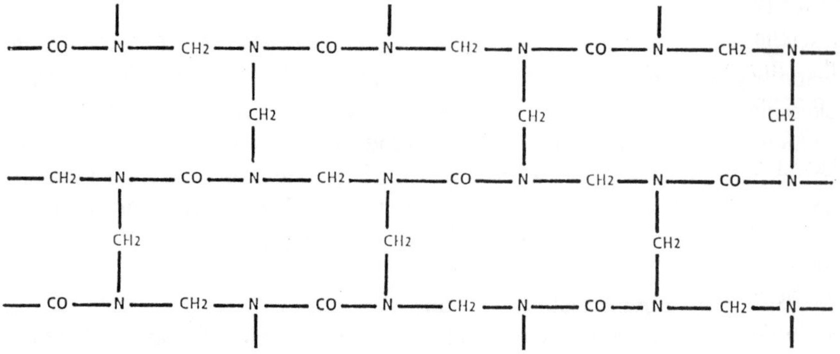

Figure 3   *Theoretical fully crosslinked dimethylol urea*

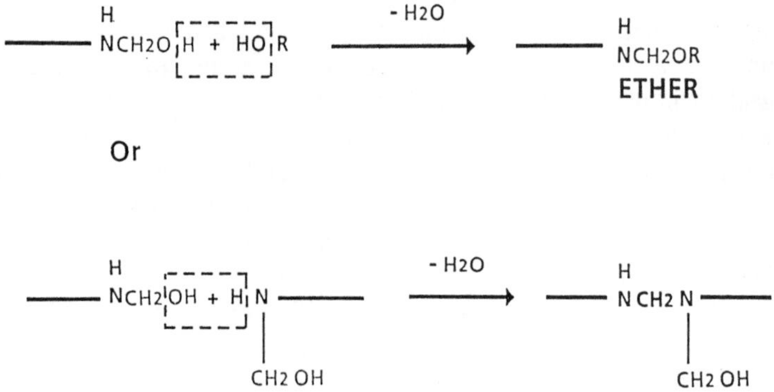

Figure 4   *Etherification reaction*

pre coated steel for subsequent fabrication, metal furniture and general industrial applications. The effect of cross-linking is to improve the hardness, toughness, chemical resistance and weathering properties of the paint. Whilst the final paint film properties are largely governed by the chemical composition of the hydroxy functional film forming polymer, the amino cross-linking component also has an important influence, largely by controlling crosslinked density.

Amino resins have developed progressively, and this has enabled then to be modified to meet the increasing demands for improved paint film performance as well as the demands which environmental legislation has imposed. This environmental legislation has led to changes from conventional low solids, solvent based paints to high solids systems, and more recently water based paints. However, in terms of the mechanism of cross-linking of amino resins, there is little difference irrespective of whether the system is solvent or water based. Before considering the cross-linking reactions in more detail it is instructive to examine the composition of amino resins and determine how the composition influences properties.

COMPOSITION OF AMINO RESINS

The apparent versatility of melamine resins is a consequence of the high functionality of the molecule, as FIGURE 6 shows. Each of the active hydrogen atoms can be unreacted, methylolated or etherified and dimers, trimers, etc can be formed by condensation. This means that the side groups shown in FIGURE 7 are theoretically possible.

A typical structure for a butylated melamine resin molecule is illustrated in FIGURE 8. It should however be remembered that a resin will consist of a mixture of molecules of different chemical compositions and chemical analysis of a resin sample will only give an average composition. It is apparent that the melamine resin has very high functionality and in combination with hydroxy functional film forming polymer a variety of reactions may occur. As already noted, the reaction kinetics are not fully understood or elucidated, but some of the possible reactions proposed by U.S workers[1] are shown in FIGURE 9.

These are given purely to indicate some of the possible reactions and are not claimed to be definitive. A curing stoving paint has been described as "seething with chemistry". The reactions shown can be divided into 3 general categories; firstly co-condensation reactions between the amino resin and the polymer, resulting in cross-linking of the system. Secondly, self condensation, or homo-polymerisation of the melamine resin. Thirdly depolymerisation, or

OH
NH CH2 O H
CO
-2H2O
NH CH2O H
OH
NH CH2O
CO
NH CH2O

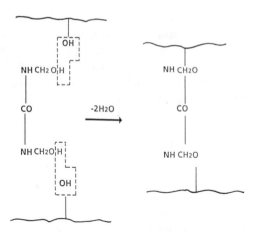

Figure 5 *Crosslinking of hydroxy functional polymer*

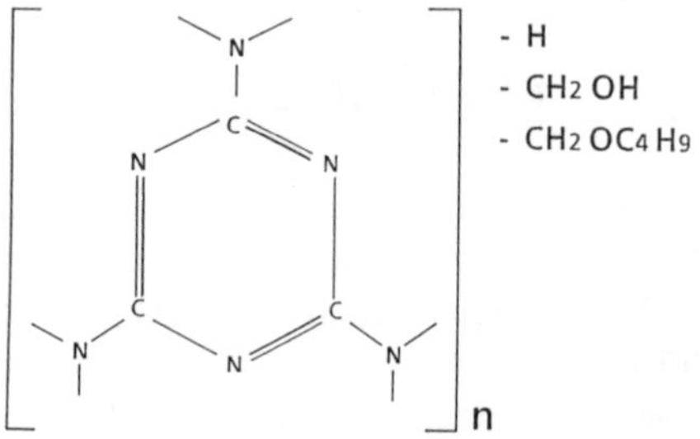

N
C
N
N
C
N
C
N
N
- H
- CH2 OH
- CH2 OC4 H9
n

Figure 6 *Possible structures of melamine resins*

—N H H
—N CH2OH H
—N CH2OR H
—N CH2OH CH2OH
—N CH2OR CH2OH
—N CH2OR CH2OR

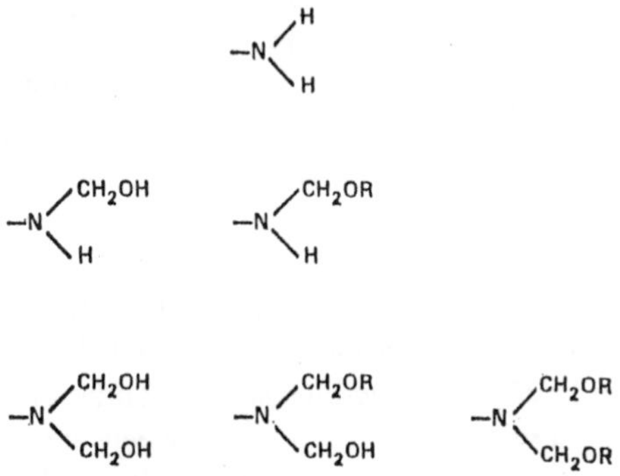

Figure 7 *Functional groups on melamine resin*

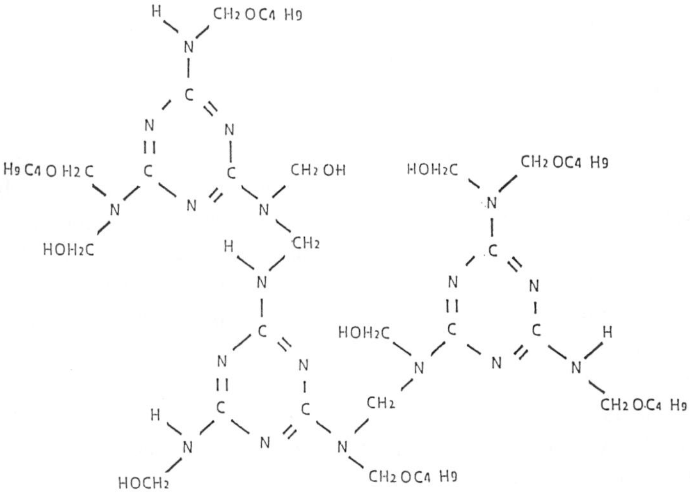

Figure 8 *Typical butylated melamine resin*

I

$$>NCH_2OH + ROH \rightleftharpoons NCH_2OR + H_2O$$

$$>NCH_2OCH_3 + ROH \rightleftharpoons NCH_2OR + CH_3OH$$

II

$$2>NCH_2OCH_3 + H_2O \longrightarrow >NCH_2N< + HCHO + 2\,CH_3OH$$

$$>NCH_2OCH_3 + >NH \longrightarrow >NCH_2N< + CH_3OH$$

$$2>NCH_2OCH_3 \rightleftharpoons >NCH_2N< + CH_3OCH_2OCH_3$$

$$2>NCH_2OH \rightleftharpoons >NCH_2OCH_2N< + H_2O$$

$$>NCH_2OCH_3 + >NCH_2OH \rightleftharpoons >NCH_2OCH_2N< + CH_3OH$$

$$>NCH_2OH + >NH \longrightarrow >NCH_2N< + H_2O$$

III

$$>NCH_2OCH_3 + HOH \rightleftharpoons >NCH_2OH + CH_3OH$$

$$>NCH_2OH \rightleftharpoons NH + HCHO$$

Figure 9 *Some reactions of melamine resin with hydroxy polymers*

degradation of the melamine resin can occur. Under more extreme conditions further depolymerisation of the system can occur through alcoholysis or hydrolysis of ester links in the film forming polymer. The phenonomen of reverse cure is observed in some applications.

Investigations[2] lead to the conclusion that the rate of co-condensation is substantially faster than self-condensation and the assumption is that co-condensation predominates in the early stages of the curing process, with self condensation taking over as the temperature rises. More recent investigation[3] has shown that all the self condensation reactions are reversible and this is compatible with the theory that a dynamic equilibrium exists which results in the relieving of stresses that may have been set up in the early stages of reaction. It has also been speculated that as condensation proceeds, phase separation of the MF resin may occur as it becomes incompatible with the increasing molecular weight polymer. This can increase the possibility of further self condensation.

Self condensation reactions generally contribute to surface hardness and stain resistance.   However they also produce brittle films with lower solvent resistance, durability, flexibility and adhesion.

To overcome the above film property disadvantages it is necessary to formulate melamine resins with a reduced tendency to self condense, and this can be achieved by increasing the degree of butylation, which blocks off the more reactive methylol groups. The consequence of doing this however is to reduce the reactivity of the resin to an unacceptable level. The only way to achieve a satisfactory compromise is to etherify the resin with methanol rather than butanol.

METHYLATED AMINO RESINS

A series of methylated resins analogous to butylated materials is available, but the most interesting of the methylated resins is the fully methylolated, fully methylated product, hexamethyl ether of hexamethylol melamine, also known as hexamethoxyl methyl melamine, or simple HMMM. Its structure is shown in FIGURE 10. The resin possesses many interesting physical as well as chemical properties when compared to conventional butylated resin and its use is continually expanding at the expense of butylated resin. HMMM a pure material is a crystalline solid which melts at approximately 40°C. However, commercial products are marketed as liquid materials at 100% solids for ease of handling. Compared to butylated resins they possess approximately double the solids content at equivalent viscosity, which makes them ideal for use in high solids, low VOC paints. Additionally, HMMM is

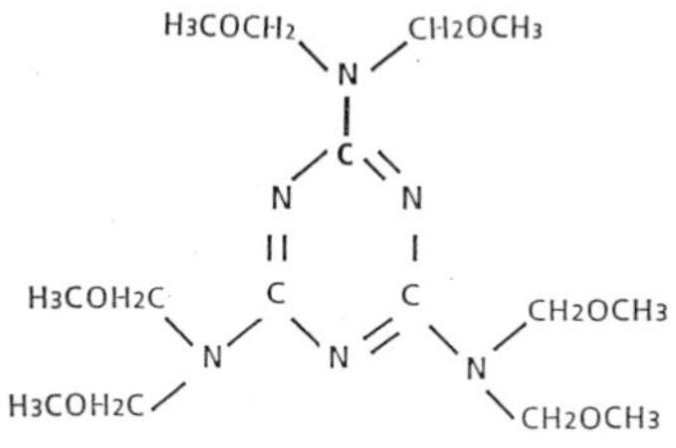

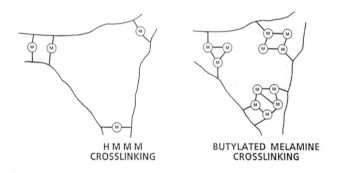

Structure of hexamethyoxymethyl melamine

$$> NCH_2OCH_3 + ROH \rightleftharpoons NCH_2OR + CH_3OH$$

Reaction with hydroxy functional polymer

Figure 10

H M M M
CROSSLINKING

BUTYLATED MELAMINE
CROSSLINKING

Figure 11   *Representation of crosslinked structures*

	BUTYLATED	METHYLATED
% SOLIDS AT 2 POISE	45	80
% FREE FORMALDEHYDE	1	0.1
% LOSS OF VOLATILE ON STOVING	29	18
% FORMALDEHYDE EMITTED ON STOVING	0.6	0.15
WATER MISCIBILITY	POOR	GOOD
SOLVENT	BUTANOL (XYLENE)	NONE

Figure 12   *Comparative properties of butylated and methylated resins*

water miscible and can be used as a crosslinker in waterborne paints, hence its importance in the current climate where strong pressure is being applied to reduce solvent emission.

In respect of its chemical reactivity, because HMMM is fully etherified, the predominant reaction which occurs during crosslinking is co-condensation, there being little tendency for self condensation to occur. FIGURE 10 shows the predominant reaction which is catalysed by a strong acid. Looked at simply, the structure of films crosslinked with butylated and HMMM resin respectively can be visualised as shown in FIGURE 11. Films crosslinked with HMMM demonstrate a) better flexibility, b) increased toughness and c) better weathering and chemical resistance compared to those produced from butylated resins. Additionally HMMM is more economical to use and emits reduced volatiles on curing, because there are fewer self condensation reactions.

CONCLUSIONS

Although the main objective has been to consider the chemistry of the preparation and reaction of amino resins it is important to emphasise that amino resins can be used satisfactorily in waterborne coatings and that in this connection methylated resins are the most suitable products, particularly HMMM. Not only are they suitable for waterborne systems, which are only one of the possible routes to environmental compliance, but they are, in their own right, more environmentally friendly than conventional butylated resins as FIGURE 12 shows.

REFERENCES

1.   *Reaction Mechanism of Melamine Resins*
     W J Blank - Journal for Coatings Technology 51 No. 656 Sept 1979

2.   *Recent Studies of the Curing of Polyester-Melamine Enamels*
     N Jones et al - Waterborne & Higher Solids coatings synopsium
     New Orleans Feb 1989

3.   *Possible Reaction Pathways for Self-condensation of Melamine Resins*
     N Jones et al - Journal of Coatings Technology 64 No. 804 Jan 19

Waterborne Two-pack Polyurethane Coatings for Industrial Applications

Annegret Bittner and Peter Ziegler

HOECHST AG, WERK KALLE-ALBERT, R&D SYNTHETIC RESINS, D-65174 WIESBADEN, GERMANY

1 INTRODUCTION

Aqueous polyurethane dispersions have gained increasing importance in a wide range of applications because of their excellent properties. These include adhesion to various substrates, resistance to chemicals, solvents and water, abrasion resistance, flexibility and toughness. Polyurethane dispersions are especially suitable for painting plastics and wood, and for coating both metallic and mineral substrates.

Conventional polyurethane resins usually contain a high proportion of volatile organic solvents, normally between 40 and 60% by weight. In 1970, the manufacturers of polyurethane resins developed processes which led to the synthesis of low-solvent or solvent-free aqueous polyurethane dispersions. The use of waterborne polyurethane coatings is growing rapidly due to the stringent controls on solvent emissions during the application process.
For example, in 1986 the German Clean Air Regulations, the so called TA Luft, suggested limits on solvent emissions for industrial coatings. Addionally CEPE, the European association of paint manufacturers suggests the reduction of solvents in automotive refinishing paints. And last but not least, the Environmental Protection Agency of the United States, established in 1963 by an act of Congress entitled "The Clean Air Act", limited the solvent content of paints. That is the so called VOC regulations, which were introduced in 1991. For example, the VOC content of automotive coatings is limited to between 250 and 340 g/l.

In the following, this paper introduces aqueous polyurethane dispersions manufactured by Hoechst for the use in two-pack polyurethane coatings for industrial applications trade named DAOTAN. First of all, comments on the basic concepts of these resins are made and distinctions between the various resins available and their fields of application. Furthermore, the present work deals with the technical development of new waterborne top coats and their properties, compared to conventional two-pack polyurethane coatings.

2 General

Aqueous polyurethane dispersions may be divided into two classes. One consists of polymers stabilized by external emulsifiers. The other can be achieved by the inclusion of hydrophilic centres in the polymer. Such centres may in principle be of one of three types

- non-ionic groups e.g. polyethylene oxide chains,
- cationic groups e.g. alkylated or protonated tertiary amines,
- anionic groups e.g. carboxylate or sulphonate groups.

The introduction of these hydrophilic groups which function as internal emulsifiers makes it possible to produce stable aqueous emulsions with an average particle size between 10 and 200 nm. „Hoechst" aqueous polyurethane dispersions are anionically stabilized.

The synthesis scheme for an aqueous, anionically stabilized polyurethane dispersion is shown in table 1. Generally, an excess of di-isocyanate is treated with a long chain, usually linear polyol, a bis-hydroxycarboxylic acid and other low-molecular-weight glycols to form an NCO-terminated prepolymer with a segmented structure.

In this polymer, the long-chain polyol units form soft segments and the urethane units built up from di-isocyanate, glycol and bis-hydroxycarboxylic acid form hard segments. Neutralization of the acid groups with amines yields the anionic hydrophilic centres which are essential for the dispersion process and subsequent stabilization of the resin in the aqueous phase.The prepolymers may be further functionalized before the neutralization step. One method of doing so is to treat the NCO groups with special reagents. Suitable neutralizing agents in this case are amines such as dimethyl aminoethanol or ammonia.

Table 1: Synthesis scheme for aqueous polyurethane dispersions

aqueous OH-functional polyurethane dispersion

3 Classification and fields of application

The polyurethane dispersions developed by Hoechst may be divided into following groups based on the differences of their chemical structure.
1. Hydroxy-functional polyurethane dispersions
2. Self-crosslinking polyurethane dispersions
3. Polyurethane-polyurea dispersions
4. Acrylated polyurethane dispersions
5. Fatty acid modified polyurethane dispersions
Last but not least, highly carboxylated hydroxy functional polyacrylic dispersions have to be mentioned, which when crosslinked with polyisocyanates also lead to polyurethane compounds.

A complete survey of the Daotan dispersions which shows the most important structural components and technical data, solvent content and general fields of application is available. The various grades have been developed for the use in special coatings like OEM primer surfacers, plastics and wood coatings, air-drying corrosion protection primers.

In the following, this paper deals with hydroxy-functional polyurethane and polyacrylic dispersions which are among others crosslinkable with polyisocyanates and are especially suitable for the formulation of aqueous two-component coatings. The properties of the resulting coatings are very similar to the conventional solvent-based ones. The incorporation of the organophilic polyisocyanate crosslinking agents into the aqueous phase is simple. This is due to the outstanding emulsifier properties of hydroxy functional polyurethane dispersions.

Table 2: Overview of suitable resins and hardeners for waterborne two-pack polyurethane coatings

$$\text{Reaction} \quad R-OH \quad + \quad R'\,NCO \longrightarrow R' - NH - \overset{\displaystyle O}{\overset{\displaystyle \|}{C}} - O - R$$

(R): Hydroxy-functional polyurethane (A) and polyarylic (B) dispersions

Type	Solids content in %	Viscosity in mPa s	Mw	OH-value
(A) Daotan VTW 1270	40	100-1000	9000	60
(B) Macrynal VSM 2521w	42	3000-4000	30000	140

(R'): Aliphatic polyisocyanates - trimerisate of hexa-methylene diisocyanate (HDI) with biuret, isocyanate, uretdione structure

Type				% NCO
Bayhydur VPLS 2980*	100	3500	-	19,5
Bayhydur VPLS 2025	100	1000	-	22,5
Bayhydur VPLS 2032*	100	3500	-	17,2

* hydrophilic

The polymer particles have small average particle size of 10 to 200 nm and are characterised by good diffusivity. It is advised to use conventional aliphatic polyisocyanate hardeners based primarily on hexamethylene diisocyanate (HDI) with a biuret, isocyanurate and uretdione structure. The viscosities of the hardeners are important. The lower their viscosity the easier it is to incorporate them into the aqueous phase. By using medium-viscosity polyisocyanates, only small proportions of organic solvent have to be added. Table 2 summarizes some examples with respect to trade names of suitable resins and polyisocyanate hardeners.

4 Formulation details

To formulate a waterborne two pack top coat, several of the processing aids usually found in water based systems are required, for example wetting agents,

Table 3: Test Formula

Waterborne two-pack top coat for industrial application based on aqueous polyurethane dispersion cross-linked with polyisocyanate Spray coating

Component A		Parts by weight	Solids
Hydroxy-functional polyurethane dispersion	Daotan VTW 1270 40%WA(Hoechst)	22	8.8
Cosolvent	Proglyde DMM(DOW)	2.5	
Wetting agent	Surfynol 104,50% Proglyde DMM (Air Products)	1.1	0.6
Defoamer	Additol VXW 4926 (Hoechst)	0.4	
Liquid light stabilizer	Sanduvor 3212 (Sandoz)	0.8	
Titanium dioxide	Kronos Titan 2310 (Kronos)	27.1	27.1
	deion.Water	4.5	
Let down: s.a.	Daotan VTW 1270 40%WA(Hoechst)	22.3	8.9
	deion.Water	8.1	
Component B			
Water-dispersable polyisocyanate crosslinker	Bayhydur VPLS 2025, 80% Proglyde DMM (Bayer)	11.2	9.0
		100.0	54.4

defoamers, light stabilizers and cosolvents. Tests with various cosolvents showed that Proglyde DMM (DOW) is best with regards to the incorporation of the hardener and the resulting appearance of the layer coating especially the gloss.

The pigment compatibility and stability during the dispersion process of the polyurethane dispersions is excellent. The batch is adjusted to the required viscosity by distilled water. The organic solvent content is approximately 8 to 12 % in the paint at a spray viscosity of 20 s measured with a Ford cup of 4 mm. The VOC values are of between 200 and 300 g/l. The calculated quantity of the polyisocyanate hardener should have a ratio of 2 : 1 calculated on solids and is homogenised shortly before application. The mixture can be achieved by means of two-component spray equipment or manually. The pot life of a mixed batch is 3 to 4 hours. A significant increase in viscosity can not be observed, in the case of polyurethane/polyisocyanate combination the batch does not gel.

The following resistance properties of the dried layer coating indicate the pot life. Their petrol resistance was evaluated by measuring the duration of exposure in minutes. As shown in table 4 the reaction time prior to the application does not effect the solvent resistance. In the case of Daotan VTW 1270 cured by Bayhydur N 3400 the resistance decreases after 2 hours. Therefore the mixture should be applied within 2 hours and the film properties are not be reduced.

Table 4: Influence of reaction time (pot-life) prior to application on the resistance properties (petrol)

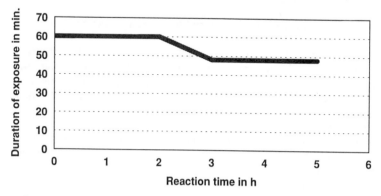

Daotan VTW 1270 cured with Bayhydur N 3400

Dry film thickness:	40µm		Drying conditions:	30 minutes at 80°C
Substrate:	glass plates			plus 16 hours at 60°C

5 Mechanism of reaction - film formation

The mechanism of reaction during the film formation is not known in detail.It is assumed that directly after mixing of the two components and subsequent application, polyol and polyisocyanate droplets are distributed evenly in the layer coating. Owing to the quick evaporation of the water and the small proportion of organic solvents, the particles are condensed into an increasingly small volume. As result, the larger polyisocyanate spheres are surrounded by numerous small polyol droplets as shown in table 5. As soon as contact is made in the surrounding area the particles diffuse. The curing reaction of hydroxyl and NCO-groups takes place in the classic manner, this means that urethane and urea groups are formed, but mainly urethane groups. Several experiments carried out by Kubitza[1] confirm the diffusion theory. Investigations of the films by differential scanning calorimetry (DSC) led to the conclusion that the layer coating has a uniform, narrow glass transition temperature just one hour after application. Furthermore, IR differential spectra show a higher mass content of urethane groups than urea.

The reaction scheme is outlined in table 6. In the first stage, the isocyanate reacts with water forming respective carbamic acid derivative which immediately decomposes again into primary amine and carbon dioxide. The resulting amine groups are able to react with further isocyanate groups in the nascent state forming N-substituted polyureas. However, there are indications that the NCO groups react mainly with the hydroxyl groups of the polymer chains in the aqueous phase. Also the reaction of NCO with water leads to the generation of carbon dioxide. The film properties are influenced by the polyurea network as well as the urethane network. Since a small proportion of the NCO groups already react with water before application to the substrate,it is advisable to use an -NCO/-OH ratio higher than 1. As a result, high quantities of free hydroxyl groups are avoided in the film.

6 Property comparison waterborne us solvent-borne coatings

In table 7 properties of films from two-pack waterborne polyurethane were compared to those films from conventional two-pack solvent-borne formulations. The coatings based on waterborne polyurethane dispersions contain very little solvent compared to the conventional systems. VOC values of 300 g/l in spite of > 500 g/l are achieved. Excellent gloss, hardness, adhesion to steel substrates and flexibility are provided by ambient cure conditions. The pot life is limited to 2 to 3 hours whereas conventional systems usually have 8 hours. In the case of polyurethane resin as mentioned above the short pot life cannot be detected by the increase of viscosity or gelation as usual. The polyacrylic dispersion leads to a visible viscosity increase.

Because of the reaction of NCO with water the occurence of reactive bubbles is possible if high film thicknesses are applied. At a dry film thickness of 60 to 70 microns an excellent appearance of the coating surface is achieved. Effective degassing of the forming film should be enabled by optimizing the formulation

Table 5:   AQUEOUS TWO-PACK PU-SYSTEMS-
MECHANISM OF REACTION-FILM FORMATION

- directly after mixing polyol and polyisocyanate droplets are statistically distributed

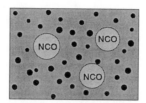

- the smaller polyol droplets arrange around the larger polyisocyanate spheres

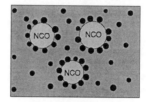

- the particles diffuse into each other

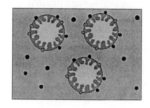

- urethane and urea groups are formed (crosslinking)

Table 6: Possible reactions

$$R-NCO \; + \; R-OH \; \longrightarrow \; R-NH-COOR$$
Urethane

$$R-NCO \; + \; H_2O \; \longrightarrow \; [\,R-NH-COOH\,]$$
Carbamic acid

$$[\,R-NH-COOH\,] \; \longrightarrow \; R-NH_2 \; + \; CO_2$$

$$R-NCO \; + \; R-NH_2 \; \longrightarrow \; R-NH-\overset{\displaystyle O}{\overset{\|}{C}}-NH-R$$

Table 7:   Property comparison of waterborne
vs. solvent-borne two-pack polyurethane coatings

| | water-borne | | solvent-borne |
	A*	B	C
% OH of polyol	60	135	80
% NCO of polyisocyanate	22	22	16,5
NCO/OH binder?	1,5:1	1,5:1	1:1
Solvent contents in %	5 - 8	9 - 12	40 - 60
Gloss 60°/20°	92/82	85/67	95/87
Pendulum hardness (s) König, force dried at 60°C/16 h	120	160	170
Adhesion Cross cut test (GT)	0	0	0
Flexibility Erichsen deep drawn test (mm)	10	9	8,5
Pot life (h)	3	2	> 8

*A   Aqueous hydroxy-functional polyurethane dispersion crosslinked with HDI trimerisate
B   Aqueous hydroxy-functional polyacrylic dispersions crosslinked with HDI
B   Conventional polyacrylic resin crosslinked with HDI

i. e. using specially chosen solvents and by taking care of the cure conditions in general and the humidity in particular.

6.1 Solvent and chemical resistance

The solvent and chemical resistance of the coatings were also studied. The resistance of the respective film properties are affected by the cure conditions, that means temperature as well as humidity. Studies of the effects of humidity on the pigmented films showed that at somewhat higher cure temperatures the

Table 8: Solvent resistance of waterborne two-pack polyurethane coatings (pigmented)

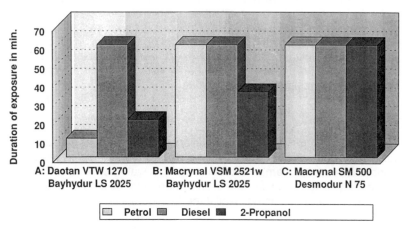

Duration of exposure in min.

A: Daotan VTW 1270   B: Macrynal VSM 2521w   C: Macrynal SM 500
   Bayhydur LS 2025   Bayhydur LS 2025   Desmodur N 75

☐ Petrol ▨ Diesel ■ 2-Propanol

Table 9: Corrosion resistance of waterborne two-pack top coatings onto waterborne epoxy primer

	A	B	C
• Salt spray resistance, 500 hrs	good	moderate	moderate
• Humidity resistance, 500 hrs	good	good	good

Weathering resistance of waterborne two-pack clear coats[)]

	A	B	C
• Xenontest 1200, 4000 hrs			
Retention of gloss in % (Reflectometer value - 60°)	80	under test	75[)]
• Florida exposure	←	under test	→

*) with light stabilizer Sanduvor 3212

2) Macrynal SM 510 n/Desmodur N with light stabilizer

negative effect of high humidity was overcome. Therefore, testing of solvent and chemical resistance was carried out under controlled curing conditions of 30 minutes at 80 °C plus 16 hours at 60 °C. In table 8 the duration of time in which the layer coatings are not attacked by petrol, diesel and 2-propanol is shown. In the case of petrol, the result of resin B is comparable to resin C - the conventional two-pack coating -whereas resin A shows a third of the time. Nevertheless, the solvent resistance showed excellent results. It has to be noted that the alkaline and acid resistance of resin B leads to comparable test results to the conventional two-pack coating.

6.2 Corrosion and weather resistance

Final results regarding the corrosion resistance and the weatherability of these new resins are not available yet. However, accelerated corrosion tests like salt spray and the humidity chamber exposures of the waterborne coatings indicate good properties. The tests were done on waterborne epoxy primers with a dry film thickness of 40 to 50 microns whilst the top coats also have 50 microns (see table 9).

The first accelerated weathering results of clear coats show promising gloss retention after 4000 hours exposed to Xenon 1200. Tests with pigmented waterborne coatings are going on presently. It is assumed that the weatherability of the new developed resins is comparable to conventional aliphatic polyurethane coatings because of their very similar chemical structure.

7 Conclusion

The need for enviromentally friendly coating systems becomes more important with the restrictions on emission of solvents.
The newly developed waterborne two-pack polyurethane coatings based on polyurethane or polyacrylic dispersions crosslinked with polyisocyanates are the choice if highest performances of the coating are expected.

References

1 W. Kubitza, Jocca 9 (1992), page 340 - 347

Coatings and Pollution Legislation

Coatings and the Development of Air Pollution Legislation

Simon T. Smith

AIR QUALITY DIVISION, DEPARTMENT OF THE ENVIRONMENT, 2 MARSHAM STREET, LONDON SW1P, UK

1 BACKGROUND

There has been an upsurge in interest in recent years concerning the pollution potential arising from the use of solvent borne coating materials in industry. Traditionally those concerns have centred around the perception of odour by the general public around the users premises, and paint particle overspray affecting people's property. Whilst these issues remain important, the air pollution debate has moved on to centre on the emission of volatile organic compounds (VOC's). Concern is often expressed about the potential health effects of VOC emissions. Arguably a greater concern, however, is that VOC's are a precursor of ozone in the troposphere and it is this pollutant which most frequently approaches or exceeds levels at which deleterious effects on human health may be observed[1,2].

2 SOURCES OF VOC EMISSION

Total man-made VOC emissions in the UK are estimated to be approximately 2,974,600 tonnes per annum. Of this overall emission solvent use accounts for 734,700 tonnes per annum, of which the painting and printing industry account for 261,300 and 41,300 tonnes per annum respectively. Whilst this may seem a relatively small proportion of the whole, the plethora of sources of VOC emissions are such that solvent use is the second largest category of man-made emission and the painting industry is by far the largest individual sector within solvent use.

A summary of the UK's VOC inventory is given in Appendix 1[3].

3 TRADITIONAL LEGISLATION

The first stage of legislative development has been related to nuisance. This seeks to protect people who are living close to the source of emission from damage to their property, acute ill health and prolonged, unpleasant sensory perception.

There is a large body of case law dealing with various aspects of nuisance[4]. In the case of Walter v Selfe (1851)(4 De G & Sm.315) nuisance was defined as:

"an inconvenience materially interfering with the ordinary comfort, physically, of human existence, not merely according to elegant or dainty modes of living, but according to plain and sober and simple notions amongst English people."

The legislation in the UK was contained within several disparate pieces of legislation until 1990. Part III of the Environmental Protection Act 1990, however, introduced a single legislative provision for nuisance and gave a single administrative enforcement procedure[5]. The problems most commonly encountered by local authority environmental health officers in dealing with public complaints about coatings users are related to the perception of odour and damage to property arising from paint particle fouling.

Complaints to local authorities about odour from industrial premises run at approximately 9,500 to 10,000 per annum[6], and relate to roughly 2,000 premises each year. In the year 1992/3, however, which is the latest for which figures are available, 13,689 complaints were made about 5,156 premises[7]. Few of these complaints result in successful prosecution, probably because either the problem is resolved satisfactorily prior to this action being taken, or because the complaint is not justified. There were 13 convictions in 1992/3.

Complaints of fallout are not documented but tend to result in the service of 100-200 abatement notices being served annually.

It is notable that this sort of legislation had local effects only and did not seek to control emissions which operate on a regional or transnational basis. It was also unpopular with industry whose required standards of environmental performance were uncertain and depended on their juxtaposition with potential complainants.

4 LEGISLATIVE CONTROL AT SOURCE

4.1   The UK

The second phase of legislative development in the UK came with Part I of the Environmental Protection Act 1990[5]. This introduced coatings users to the concept of control at source for the first time. Generally speaking, any coating process which uses in excess of 5 tonnes of organic solvent per annum is subject to this form of control. Vehicle refinishers are covered in excess of 1 tonne per annum and a couple of printing processes do not come under control unless they exceed 25 tonnes per annum. The calculation of solvent "use" is defined in the Environmental Protection (Prescribed Processes and Substances) Regulations 1991, SI472[8] as amended by the Environmental Protection (Prescribed Processes and Substances Etc)(Amendment) Regulations 1994, SI1271[9] as:

"a.   the total input of organic solvents into the process, including both solvents contained in coating materials and solvents used for cleaning or other purposes; less

b.   any organic solvents that are removed from the process for re-use or for recovery for re-use."

Although not detailed in the legislation itself the anticipated standards of control are laid down in guidance issued by the Secretary of State under the Act. Those relevant to coating are listed in Appendix 2. Each guidance note is written for a particular type of coating process, except PG6/23: Coating of Metal and Plastic, which is intended to be a catch-all for the miscellaneous general coating processes for which no specific note exists.

As regards the traditional problems odour is subject to stricter criteria than previously, and the paint particulate in the exhaust gases is limited in its concentration to a level which implies filtration in most cases. With VOC's two routes to compliance are recognised. The first of these is to use abatement equipment to capture VOC emissions arising from the use of conventional solvent borne paints. Depending on the nature of the process and of the emission, the VOC's may be recovered or destroyed. The other route to compliance would involve moving from conventional solvent borne coatings to coatings which contain less organic solvent. The solvent content of different types of coating is specified in the end use standards and implies the use of high solids or water borne coating technology. Alternatively the coatings user could go one step further and, if process parameters allowed, move to powder coating technology.

Although there are other important aspects of control such as containment and solvent auditing, the final important aspect is that of the application equipment used to get the coating onto the substrate. Criteria are laid down for transfer efficient means of applying paint[10]. HVLP, Air Assisted Airless, Electrostatic, Centrifugal and, under some circumstances, Airless spray application systems are acceptable but conventional air assisted equipment is considered too wasteful. The means of expression does not preclude the possibility of non-sprayed applications such as roller, curtain or dip coating. It should lead to less paint overspray and therefore less waste being created, which in itself should reduce the mass of VOC emitted from a process. This may well be an important consideration for formulators since the coatings they develop from now on will have to work with spray application equipment other than conventional air assisted spray guns. Ironically these controls militate in favour of the second oldest form of coating application equipment - the paintbrush.

The change from solvent borne to water borne coating technology may worsen, or at least alter the impact of coating manufacture and use on the water environment. The Department of the Environment is currently considering the impact of VOC reduction measures and what may consequently be needed in terms of control over discharges to water.

4.2   In Europe

The EC has also seen the need to move on this issue and is in the process of drafting a Directive concerning air emissions of solvents from industry. Unsurprisingly the scope of this draft Directive is similar to that in UK legislation[11].

The draft Directive also envisages two routes to compliance, one based on abatement and the other based on coating re-formulation. There is however an important difference, since the European Commission considered that Community Law would not allow them to propose a Directive with low solvent coatings specifications under an environmental treaty base. They have therefore developed a form of words which expresses the reduction in terms which uses the abatement route as a benchmark. An example is given in Appendix 3, and its practical corrollary in Appendix 4[12]. This linkage may be especially important in the future. It implies that policy makers may expect low VOC coatings to achieve emissions at least as low as if abatement had been applied to processes using conventional solvent borne coatings.

Also within the draft Directive is a provision concerning National Programmes. This allows member states to depart from the detail of the Directive provided there is a national programme of VOC reduction in place, and provided it achieves VOC emission reductions in the sectors affected by the Directive at least as great as if the detailed Directive had been implemented in full. This approach is claimed to give Member States greater flexibility in achieving emission reductions and therefore to take account of the principle of subsidiarity. The inclusion of an article allowing the use of economic instruments may allow a wide range of reduction measures to be examined, particularly where no command and control system is currently in place.

Most countries in the wider Europe will have to consider their VOC reduction programmes against the backdrop of the UNECE VOC Protocol; the Geneva Protocol[13]. This Protocol has a number of aspects but four in particular are worthy of note. The most immediate amongst these is the requirement to reduce VOC emissions by 30% by 1999, using 1988 as a baseline. The UK has published a VOC strategy[1] which predicts that current controls will lead to a reduction in emissions of 36%.

Also within the Protocol is a commitment to enter into negotiations for a further percentage reduction in VOC emissions. This is potentially significant, as will become clear later in the paper.

All contracting countries have to make a commitment to apply best available technologies that are economically feasible to new and existing stationary sources.

The other interesting aspect of the Protocol is an invitation to discriminate between VOC's which are effective photochemical oxidants and those which have relatively little effect. A great deal of development work has been taking place to rank VOC's according to their effectiveness, particularly in the UK[14]. The concept of Photochemical Ozone Creation Potential (POCP) is seen as a useful means of differentiating between VOC's on their propensity to create ozone and this index, multiplied by the mass of VOC release probably gives the best indication of the overall ozone creating effect of the VOC release. When last reviewed it was thought that it was not yet robust enough to be used as a law enforcement tool, and may have disadvantages in complexity and cost especially where monitoring was concerned.

5 A RETURN TO EFFECTS BASED LEGISLATION

The third type of legislation is a return to effects based control. The European Community set a Directive 92/72 EEC in 1992 concerning ozone[15]. It set a population information threshold and a population warning threshold. This is likely to be superseded in time by a new Directive. An independent body in the UK has also been addressing acceptability criteria for levels of ozone in the air, and reported earlier this year[16].

The European Commission has recently adopted a proposal for a Framework Directive on Ambient Air Quality Assessment and Management[17].

In its current form this would seek to set health based limit values and alert thresholds for a range of pollutants. The aim is to attain ambient air concentrations which are under putative no-adverse-effect levels and to provide information to the public. Ozone is one of those identified as being a priority substance requiring the setting of standards before the end of 1996. Where ambient levels exceed the limit value plus a permitted margin of exceedance, the Member State is to produce a plan or programme to result in the limit value being attained within a fixed time period.

The third report of the UK Photochemical Oxidants Review Group indicates that the current range of VOC and NO_x controls will be insufficient to prevent exceedances in the UK of the EC Ozone Directive population information threshold[18]. Furthermore, recent work by Derwent and Davies[19] indicates that NO_x reductions of the order of 80% or VOC reductions of the order of 70% would be required to attain that objective.

The EC has a useful role to play in the promulgation of effects based legislation. It should enable a more efficient targeting of control on those sources which give rise to pollutant effects on the most people or on the most receptor sensitive areas. An EC framework is desirable in that it should lead to common means of assessing and reporting air quality. Additionally, the air concentrations which create adverse health effects should not vary from Member State to Member State.

Furthermore, it should be recognised that effects based legislation is unlikely to be enough in isolation where some transboundary pollutants are concerned. Elevated ozone levels in southern England during episodes are significantly contributed to by precursor emissions and ozone imported from continental Europe[20]. The UK of itself can only do so much to reduce ozone levels on its territory, the remainder of the effort having to come from elsewhere. There is an argument, therefore, for EC technology based Directives in these cases to reduce precursor emissions from the source countries, who of themselves are unlikely to see the greatest benefit.

6 OTHER RELEVANT LEGISLATIVE DEVELOPMENTS

There are also other related developments which are not solely concerned with air pollution but may impinge on legislative and other controls.

6.1 Integration of Pollution Control Regimes

The European Commission has proposed a Directive on Integrated Pollution Prevention and Control[21]. This would require a wide range of industry to be issued with a single permit covering discharges to air, land and water. It is in many ways similar to the UK Integrated Pollution Control approach. Whilst the only coating process covered by UK IPC is the application of tributyltin and triphenyltin coatings, the proposed IPPC Directive has a far wider scope in the coatings sector. In its current form it includes all coatings use and manufacture involving the use of more than 200kg solvent per hour, the pharmaceutical industry, textile treatment, coating of steel and vegetable oil extraction among others. No consensus has yet been reached on the appropriate scope for this Directive and negotiations are still taking place.

Another significant issue is the interface between the Integrated Control and Organic Solvent Emissions Directives since the emissions limits in the latter would be binding on the processes subject to integrated control. Within the concept of integrated pollution control these limits may not represent the best solution, taking all environmental media into account. Fixed limits applied to releases to individual environmental media may, however, restrict the flexibility necessary to arrive at a control package which minimises impacts on the environment taken as a whole and which represents the best overall solution.

Another area of debate which may influence the final shape of the Directive and its implementation in Member States is in the apparent difference in perceptions as to what an integrated approach to control actually comprises. This could comprise one enforcing authority fully considering the tradability between environmental media or merely a single permit issued by the existing regulatory structures with better co-ordination between them.

6.2   Control from an Unexpected Quarter

The Paris Commission is an intergovernmental organisation which administers a convention designed to protect the marine environment in the North East Atlantic. This includes emissions directly to those waters, and indirectly via inland waters or the air. One of its ongoing programmes of work is to promulgate environmental standards to protect that marine environment. It is currently considering a report from the Netherlands which examines emissions from paint products to the aquatic environment[22].

The current draft recommendations being considered by the organisation include:

a.   the application of Best Available Techniques in paint production and use
b.   effective systems for collection and processing/disposal of paint residues
c.   eco-labelling and classification
d.   strategies to reduce hazardous components by controls on marketing and use.
e.   promulgation of Best Environmental Practice in the use of decorative paint
f.   identification, quantification and risk assessment of hazardous paint components
g.   development of less adversely impacting alternatives

The recommendations have yet to be adopted.

6.3 Labelling

In addition to industrial sources coatings are also used in many thousands of other locations, notably on an ad-hoc basis for domestic decoration/protection. Whilst these sources are insignificant individually, they aggregate to an important difuse source. It would not be practical or desirable to implement a command and control system for these sources so alternative strategies have to be found.

VOC emissions may be effected by controlling the solvent content of the coating itself. Whilst such controls are not being drawn up at present the European Community is considering issuing eco-labelling criteria for decorative indoor paints and varnishes[23]. A Commission decision is imminent, and seems likely to allow the eco-label to be fixed to products with a VOC content lower or equal to 30g/l (minus water) and a volatile aromatic hydrocarbon content lower or equal to 0.5% for paints with a specular gloss below or equal to 40 units at $\alpha = 60°$. For paints with a specular gloss above 40 units at $\alpha = 60°$, and for varnishes the figures are likely to be 250g/l(minus water) and 1.5% respectively.

7 CONCLUSIONS

Five tentative conclusions can be drawn from the development of air pollution legislation in recent years and from the discussions which are taking place.

1.   The UK has recently set stringent controls over VOC emissions from a range of industrial coating processes, allowing low solvent coatings as an alternative to abatement. The specification of these low solvent coatings is likely to become more demanding as paint technology world-wide develops.

2.   In its draft proposals the European Commission has made a linkage between the potential emission reductions achievable by abatement and those achievable by low solvent coatings. This is likely to be an attractive philosophy for policy makers.

3.   Given the recent work on air quality standards and the reductions which would be needed to achieve those levels, it is possible that future international agreements brokered by the UNECE will require more aggressive reductions in VOC emissions.

4.   There are likely to be greater moves to the integration of the disparate pollution control regimes.

5.   Coating formulators need to be alive to the squeeze on solvents, potentially hazardous components of coatings and on transfer efficiency of coatings.

APPENDIX 1.[3]   **SUMMARY OF ANTHROPOGENIC NON-METHANE VOC EMISSION ESTIMATES BY SOURCE (kt/a).**

SOURCE	EMISSION kt/a
SOLVENT USE	
Painting industry	261.3
Printing industry	41.3
Metal cleaning	42.6
Dry cleaning	11.0
Adhesives	58.0
Pharmaceuticals	40.0
Aerosol consumer products	85.6
Non aerosol consumer products	73.0
Agrochemicals	39.0
Seed oil extraction	10.0
Leather industry	2.3
Other solvent use	70.6
TOTAL SOLVENT USE	734.7
OIL INDUSTRY	
Crude oil production	92.0
Crude oil distribution	107.0
Oil refineries	180.0
Refinery product distribution	128.0
Other oil industry	25.0
TOTAL OIL INDUSTRY	532.0
CHEMICAL INDUSTRY	200.0
STATIONARY COMBUSTION	
Domestic combustion - coal	59.8
Domestic combustion - other fuel	14.1
Industrial combustion	5.6
Electricity generation	6.0
Fuel manufacture	?
Other combustion	0.9
TOTAL STATIONARY COMBUSTION	86.4

APPENDIX 1. [3]   **SUMMARY OF ANTHROPOGENIC NON-METHANE VOC EMISSION ESTIMATES BY SOURCE (kt/a). ...continued**

SOURCE	EMISSION kt/a
FOOD INDUSTRY	
Bread baking	7.4
Alcoholic beverages	44.0
Animal by-products	0.4
Other food processing	22.2
TOTAL FOOD INDUSTRY	74.7
IRON & STEEL INDUSTRY	9.1
WASTE DISPOSAL	
Landfill	10.0
Incineration	3.2
Straw/stubble burning	15.4
Sewage treatment	0.4
TOTAL WASTE DISPOSAL	29.0
AGRICULTURE	
Animal husbandry	56.0
Arable farming	241.5
TOTAL AGRICULTURE	297.5
MISCELLANEOUS	39.9
TOTAL STATIONARY SOURCES	2002.6
MOBILE SOURCES	
Petrol engines - exhausts	644.0
Petrol engines - evaporation	137.0
Diesel engines	167.0
Other transport	24.0
TOTAL MOBILE SOURCES	972.0
TOTAL ALL SOURCES	2974.6

APPENDIX 2

PG 6/7	Printing and Coating of Metal Packaging
PG 6/8	Textile and Fabric Coating and Finishing Processes
PG 6/9	Manufacture of Coating Powder
PG 6/10	Coating Manufacturing Processes
PG 6/11	Manufacture of Printing Ink
PG 6/13	Coil Coating Processes
PG 6/14	Film Coating Processes
PG 6/15	Coating in Drum Manufacturing and Reconditioning Processes
PG 6/16	Printworks
PG 6/17	Printing of Flexible Packaging
PG 6/18	Paper Coating
PG 6/20	Paint Application in Vehicle Manufacturing
PG 6/22	Leather Finishing Processes
PG 6/23	Coating of Metal and Plastic
PG 6/31	Powder Coating (including sheradizing)
PG 6/32	Adhesive Coating
PG 6/33	Wood Coating
PG 6/34	Respraying of Road Vehicles
PG 6/40*	Coating and Recoating of Rail Vehicles
PG 6/41*	Coating and Recoating of Aircraft

*   To be published shortly

APPENDIX 3 [11,12]

EXAMPLE OF A NON-ABATEMENT ROUTE TO COMPLIANCE IN THE DRAFT EC SOLVENTS DIRECTIVE

Annex IX : Vehicle Refinishing Processes

5. Exemption. Member states may exempt the operator from the compliance with the emission limit values given in items 2 and 3 if the operator forwards an emission reduction plan which includes in particular decreases on the average solvent content of the total input and/or improvements of the application techniques and which results in a reduction of annual emissions, complying with the following objectives:

Time period in years after the Directive comes into force for		Maximum allowed total annual emissions of the installations due to the implementation of the reduction plan, expressed in percent of total annual reference emissions to be achieved after the time periods indicated on the left at the latest
new installations	existing installations [+]	
2	4	100

[+]   including the time period mentioned under item 1.4

The total annual reference emissions of the installation are equal to the maximum emissions which would occur if the fugitive emission limit value and the abatement of stack emissions with an efficiency of [85] percent were applied to the same installation, assuming an average weight percentage of organic solvents in the total input of [75].

The total emissions are equal to all outputs defined in Annex I, but 05, 07 and 08. The total input is equal to the sum of the weight of all preparations taken into account to determine 1_1, 1_2 and 1_3, as defined in Annex I, and all preparations which are free of organic solvents and which replace directly and for the same purpose preparations which contain organic solvents, assuming the installation is operated at nominal capacity.

The plan shall be revised if the nominal capacity is increased by more than [25] percent.

For an installation which runs under the exemption given above, solvent management plans of type B have to be established, starting with the first year of implementation of the emission reduction plan.

An installation, which runs under the exemption given above, may be exempted from the application of Article 9, paragraph 2 if such a measure entails no significant worsening of the objective concerning the accuracy laid down in Annex I.

APPENDIX 4 [11,12]

CALCULATION OF THE REDUCTION EQUIVALENT TO ABATEMENT IN ANNEX IX OF THE DRAFT EC SOLVENTS DIRECTIVE

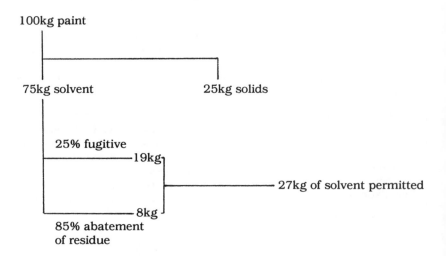

This equates to a range of products of 53% solvent by weight to achieve the same final coated weight.

REFERENCES

1.   Department of the Environment, "Reducing Emissions of Volatile Organic Compounds (VOC's) and levels of Ground Level Ozone: A UK Strategy", 1993

2.   Warren Spring Laboratory,   "Air Pollution in the UK: 1992/3", LR1000(AP), J S Bower et al, 1994.

3.   Warren Spring Laboratory, "Emissions of Volatile Organic Compounds from Stationary Sources in the UK", LR990, N R Passant, 1993.

4.   C N Penn, "Noise Control", Shaw & Sons Limited, London, 1979.

5.   "Environmental Protection Act 1990", Chapter 43, HMSO, London, 1990.

6.   Institution of Environmental Health Officers, "Environmental Health Report 1987 - 1990", 1991.

7.	Institution of Environmental Health Officers, Environmental Health Journal, 1994, **102**, p217.

8.	"The Environmental Protection (Prescribed Processes & Substances) Regulations 1991", SI 1991: No. 472, HMSO, London.

9.	"The Environmental Protection (Prescribed Processes & Substances Etc)(Amendment) Regulations 1994", SI 1994: No. 1271, HMSO, London.

10.	S Mannouch, "Deposition Efficiency of Modern Spray Equipment", Paint Research Association Seminar Proceedings, Pollution Abatement in Surface Finishing, 1994.

11.	"Draft Proposal for a Council Directive (EEC) on the limitation of the emissions of organic compounds due to the use of organic solvents in certain processes and industrial installations", European Commission, July 1994.

12.	"Solvent Emission Reduction in the Vehicle Refinishing Industry", CEPE, 1994.

13.	"Protocol to the 1979 Convention on Long Range Transboundary Air Pollution Concerning the Control of Emissions of Volatile Organic Compounds or their Transboundary Fluxes", United Nations Economic Commission for Europe, Geneva, Switzerland, 1991.

14.	R G Derwent and M E Jenkin, Atmospheric Environment, 1991, **25A**, pp1661-1678.

15.	"Council Directive 92/72/EEC on air pollution by ozone", Official Journal, L297, p1, 13/10/92.

16.	Department of the Environment Expert Panel on Air Quality Standards, "Ozone", HMSO, London, 1994.

17.	European Commission, "Explanatory Memorandum concerning the proposal for a Council Directive on Ambient Air Quality Assessment and Management", XI/860/93, Rev. 1, 1994.

18.	United Kingdom Photochemical Oxidants Review Group, "Ozone In The United Kingdom 1993", Third Report, Department of the Environment, 1993.

19.	R G Derwent and T J Davies, Atmospheric Environment, 1994, **28**, pp2039-2052.

20.	United Kingdom Photochemical Oxidants Review Group, "Ozone in the United Kingdom Interim Report", First Report, 1987.

21.	European Commission, "Proposal for a Council Directive on Integrated Pollution Prevention and Control", 14/9/93.

22.	PARCOM, "Emissions from Paint Products to the Aquatic Environment; Presented by the Netherlands", Diffchem 4/5/2-E, 1993.

23.   European Commission,   "Draft   Commission   Decision   of   1994
Establishing the Criteria for the Award of the Community Eco-Label to Indoor
Paints and Varnishes", 1994.

Additives

Foam Control Agents for Waterborne Coatings

E. C. L. van Laere

DREW INDUSTRIAL DIVISION, ASHLAND CHEMICAL COMPANY (A DIVISION OF ASHLAND OIL, INCORPORATED), TRIATHLONSTRAAT 33, NL-3078HX ROTTERDAM, THE NETHERLANDS

1 DEFOAMERS AND ANTI-FOAMS

Foam Control Agents (F.C.A.'s) can be divided into defoamers and anti-foams. Defoamers are process-additives which act to control foam during an industrial process. Anti-foams are product-additives which are used to prevent foam formation.

Quick but transient foam control is obtained when using a defoamer. More persistent foam control is achieved by utilizing an anti-foam.

Coating formulations normally make use of an anti-foam which can be added in one or two steps (grind and/or letdown). Sometimes a defoamer plus anti-foam combination is utilized.

2 INCREASING PERFORMANCE AND APPLICATION STANDARDS

Waterborne coatings are demanding more and more from F.C.A.'s due to:
- Higher performance standards such as a higher film thickness, higher gloss levels, lower V.O.C. levels and lower odour levels.
- More difficult application methods due to increased coating speeds and spraying pressures.
- Quicker and more automated production techniques which require more easy to handle, to pump and to emulsify/disperse F.C.A.'s.

3 FOAM CONTROL (AGENT) PROBLEMS

Entrapped macro-foam affects the paint production, filling off and paint film properties. One of the most difficult problems is the presence of micro-foam which isn't de-aerated from the surface film quick enough due to the presence of slow evaporating solvents. The consequence of this is the formation of pinholes in the surface film during drying.

Another often observed phenomenon is the decrease of the foam control activity with time. Surface active materials in the coating formulation are responsible for this. They are "wetting" the F.C.A. The F.C.A. is too finely dispersed and becomes too compatible with the system resulting in a loss of activity.

4 THE NATURE OF FOAM (STABILITY)

Foam is a dispersion of gas in a liquid. The presence of a surfactant in the liquid is necessary to produce a stable foam. In waterborne coatings, surfactants can be present in the form of emulsifiers (to stabilize the polymer particles), wetting agents and pigment-dispersants.

5 MECHANISMS OF FOAM CONTROL

Water soluble solvents such as methanol, ethanol or acetone may have some defoaming properties.
They work by increasing the solubility of surfactants in the system. They don't perform as anti-foams and hence their use is limited.

Water-immiscible liquids are usually more efficient anti-foams. These liquids must be able to wet the bubble surface film and to spread out across the surface film . The surface film will then become thinner until the bubbles will burst.

Mathematically this can be represented by:

$$\gamma a - \gamma m - \gamma int < 0 \quad \text{needed for wetting}$$
$$\text{and} \quad \gamma m - \gamma a - \gamma int > 0 \quad \text{needed for spreading}$$

γm = surface tension of medium
γa = surface tension of anti-foaming liquid
γint = interfacial tension

Combination of those equations implies that:
γ anti-foam $< \gamma$ medium

To 'boost' the foam control activity and persistency of the anti-foam liquid, hydrophobic particles such as hydrophobized silicas, fatty waxes and urea are used.

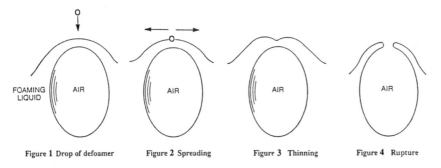

Figure 1 Drop of defoamer   Figure 2 Spreading   Figure 3 Thinning   Figure 4 Rupture

Those particles have very large hydrophobic surface areas. This causes additional repulsion between bubbles and acts as a point source for rupture (fig. 5).

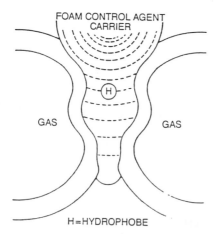

Figure 5   Multicomponent foam control agent

Emulsifiers can be added to the anti-foam formulation to increase the miscibility/compatibility of the product with the medium.

Finally, the anti-foam liquid should have a density lower than the density of the medium to be able to apply its activity at the surface of the medium.

In summary: Anti-foams need the following properties:
* a surface tension lower than that of the foaming medium
* some degree of immiscibility/incompatibility with the medium
* a density lower than the medium-density

6. FOAM CONTROL AGENT FORMULATIONS

Foam Control Agents are normally supplied as liquids.
Sometimes they are supplied as powders, produced by absorbing the liquid product on to a powder-carrier.
As discussed in the previous paragraph, F.C.A.'s are formulated out of three main groups of ingredients.

6.1. Water-immiscible liquids
Typical carriers in use are:
- Mineral oils * Paraffinic
                * Naphthenic
                * Aromatic
                * Blends
- Silicone oils (PDMS - Polydimethylsiloxane)
- Fatty acids
- Vegetable oils
- Fatty alcohols
- Silicone surfactants (Polyalkylene oxide modified PDMS)
- Polypropylene glycols

These liquids normally constitute 75-80% of the total active ingredients of the F.C.A. (Sometimes an F.C.A. is supplied as an emulsion in water, see 6.5.)
They have a large influence on the compatibility and persistency of the F.C.A. (With persistency we mean the activity of the F.C.A. after ageing of the foaming system.)
Silicone oils are the most extreme in this respect: they are quite incompatible (the higher their molecular weight and hence their viscosity, the more incompatible they are) but also quite persistent.
Alcohols and glycols are generally not very persistent and become too compatible after ageing in the system. About mineral oils can be said that the persistency and incompatibility increase from paraffinic oil, via the naphthenic oil to the aromatic oil.

Besides the 3 characteristics given in paragraph 5 these liquids might also have properties which are undesirable for the dry coating film. Besides the incompatibility which might lead to surface defects such as fish eyes and orange peel-effect, wetting may be reduced and recoatability problems (sometimes with high dosages of silicones), gloss reduction and haziness (most mineral oils) can occur.

These liquids are usually also the critical raw material regarding FDA or BGA compliance of the F.C.A. which is of importance for food-contact-coatings/printing inks.

6.2. Hydrophobes
Most commonly used hydrophobic agents are:
- Hydrophobic silica
- Fatty waxes
- Polyureas
- Metal soaps

Combinations of two different hydrophobes can be used in one formulation.
Some mineral oil based anti-foams make use of small amounts of high molecular weight silicone oil or polypropylene glycol as hydrophobes.
The type and amount of hydrophobe(s) used greatly influences the activity and persistency of the F.C.A.
The total amount of hydrophobe(s) typically lies between 5 and 10% of the total active ingredients.

6.3. Emulsifiers
Types of (non-ionic) emulsifiers used in F.C.A. formulations are:
- ethoxylated fatty acids (esters)
- ethoxylated fatty alcohols
- ethoxylated nonyl phenols
- sorbitan esters

Emulsifiers are used to make the F.C.A.'s (more) emulsifiable in the foaming system.
Easy emulsifiable products usually are quick-working but not long lasting (persistent) when compared to equivalent F.C.A.'s without emulsifier. Products without emulsifier-content are best dispersed in coating formulation during the grind stage.
So the ease of emulsification of a F.C.A. influences its initial activity and its persistency.
As with hydrophobes, sometimes emulsifier-combinations are used and the total amount varies between 5 and 10%.

Emulsifiers might influence coating properties such as water uptake and colour acceptance.

6.4. Minor components
Other components used in F.C.A. formulations are thickeners, diluents and preservatives, which are normally used to improve the properties of the wet F.C.A.'s.
The total amount of these 'additives' is normally less than 5% of the F.C.A. formulation.

6.5. Water
There is an increasing tendency to supply F.C.A.'s as an 20 to 30% emulsion in water. These emulsions offer as benefits their low viscosity (easy to pump) and their

ease of emulsification. Those properties make them suitable in automated paint production plants and in non-pigmented coating formulation. So far, types on the market are mainly emulsions of silicone oil-based F.C.A.'s although mineral oil-based emulsions are also available already.

7. FOAM CONTROL AGENT-BALANCE

As the compatibility of a Foam Control Agent decreases, its activity will normally increase.
This (in)balance can be represented as follows:

For activity a F.C.A. needs properties like:
- a lower surface tension than the foaming medium
- insoluble in the medium
- a lower density than the medium

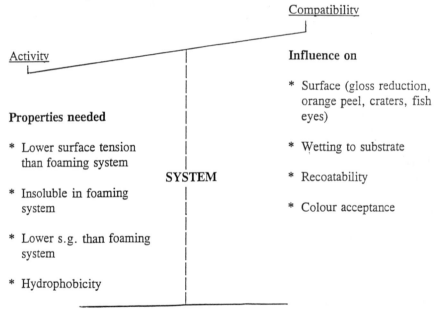

Compatibility

Activity   **Influence on**

Properties needed

* Lower surface tension
  than foaming system   * Surface (gloss reduction,
                                                          orange peel, craters, fish
                                                          eyes)

                              SYSTEM   * Wetting to substrate

* Insoluble in foaming   * Recoatability
  system
                                                       * Colour acceptance

* Lower s.g. than foaming
  system

* Hydrophobicity

The F.C.A.'s compatibility will influence properties of the Surface Coating (Medium) such as:
- orange peel-effect/craters/fish eyes/cissing
- substrate wetting
- recoatability

If the medium changes, the balance can change completely. Especially the types and amounts of surfactants and (co-)solvents in the medium have a large influence on the balance as they will affect the coatings surface tension and the solubility of the F.C.A. in the medium.

For example: a certain silicone-based product performs as a very active, slightly incompatible (some craters visible) F.C.A. in a low V.O.C.-coating. In a conventional, solvent-based, coating this product appears to perform as an anti-crawling agent without foam control activity.

The F.C.A. performance is also strongly influenced by the stage in the coating manufacturing process where the F.C.A. is added. This is especially the case during production of a pigmented formulation.

When the F.C.A. is added during the grind stage, this results in a lower activity and persistence, but a better compatibility than when the F.C.A. is added in the letdown stage. The high shear conditions and high dispersant concentration in the mill base are causes of the F.C.A. activity drop. Usually a compromise is made by adding part of the F.C.A. during the grind stage to control the (micro-)foam during production and another part (½ or ^{1}/3) in the letdown for the anti-foaming properties (persistence).

8. FOAM CONTROL AGENT TESTING

Testing of different Foam Control Agents in a certain medium means that one is looking for the F.C.A. which gives the best balance between activity and compatibility.

8.1. Activity Tests

The purpose of those tests is to generate foam/air-entrapment and to measure differences in foam control/air release.

Depending on the viscosity of the system, one can perform stir tests (high viscosity), shake tests (medium viscosity) and air-sparge tests (low viscosity).

8.1.1. Shake Test

* A paint can is filled for ± 50% with the foaming medium plus F.C.A.
* The can is shaken for 5 minutes on a 'Red Devil Paint Shaker".
* Immediately after shaking the density of the sample is measured by pouring the sample into a tarred 100 mls cylinder and recording the weight.
* The amount of entrapped air is calculated from the difference in density before and after shaking.

8.1.2. Stirrer Test

Same as shake test except for the fact that the sample is stirred (usually for 5 minutes at 2000 r.p.m.using a propeller stirrer) instead of shaken.

8.1.3. Air Sparge Test

* Pour 200 mls of the foaming system into a 1000 mls graduated cylinder.
* Add the (prediluted) F.C.A. at the required dosage.

* Place a gas dispersing tube in the cylinder and set the airflow meter at a rate of 5 ltr per minute.
* Turn on the airflow and timer.
* Record the foam heights (mls marks) after 30, 60, 120 and 300 seconds.

8.2. Application Tests

The coating/printing ink/adhesive etc. is applied by brush, roller, spray gun etc. over a certain substrate. The dry film is checked for entrapped air, craters due to broken bubbles and surface defects caused by F.C.A. incompatibility such as fisheyes, cissing, void spots.

8.2.1. Compatibility Test

- The compatibility of the F.C.A. with the system is usually also checked by applying it by a wire-wound bar or bird applicator over a glass panel.
- The dried film is checked for surface defects due to incompatibility. Entrapped micro-foam can also be seen when holding the panel against the light.

The application   and compatibility test are usually carried out with a coating (+ F.C.A.) sample which had been standing overnight.

8.3. Persistence Tests

The persistence (activity after a period of time) of a F.C.A. is determined by repeating the activity test after (heat)ageing the samples.
As a standard 14 days at 20°C or 40°C is used; as the F.C.A. activity is usually stable after that period.

Foam Control Agent selection is usually one of the last steps of the formulation work. However F.C.A.'s are normally not added as the last component of the coating-manufacturing process.
Usually they are added in two steps: partly (½ or $^2/3$) during the pigment grind step and partly during the letdown process. F.C.A.'s should be incorporated in the formulation likewise in the laboratory. As described in paragraph 7, the addition point greatly influences the F.C.A. performance.

9. FUTURE/DEVELOPMENTS

The quality of waterborne coatings is improving and they are being utilized by more sophisticated application techniques in new areas where surface defects become more apparent.
The amounts and types of (co-)solvents allowed in waterborne coating formulation are constantly decreasing.
Automated manufacturing methods are demanding stable raw materials which are easy to handle and to utilize.

These trends are requiring a new generation of raw materials. In the case of F.C.A.'s the trend is to have products which:

* Do not contribute to the (Volatile Organic Compounds) V.O.C. levels of the formulations. This implies that certain light mineral oils cannot be used anymore, as a F.C.A raw, as their vapour pressure is above 10 Pascal at 20°C. Silicones, polyglycols & vegetable oils do not present problems so far.
* Silicone-based raw materials were and are usually the most active F.C.A.'s. They will play an increasing important role in F.C.A. formulations. Polypropylene glycols are also utilized more and more because of their good air-release properties and their limited influence on gloss. Hydrophobized silicas are used to increase the activity and persistence of these products.

F.C.A.'s are supplied more and more in the emulsion form. These, low viscosity, products are easy to pump and to emulsify in the coating formulations. Drawback is that the user costs are higher than for 100% products because of the presence of water plus extra emulsifier. Therefor they are mainly successful as silicone emulsions in high cost formulations.

We can conclude that, as the quality and application techniques of waterborne coatings are developing, the need for high quality F.C.A.'s increases. Current developments are going in the direction of emulsions of silicone-silica-polyglycol based products.

An Additives Approach to Defect Elimination in Thermoplastic Waterborne Industrial Maintenance Coatings

Joel Schwartz,[1] Stephen V. Bogar,[1] and William R. Dougherty[2]

[1]AIR PRODUCTS AND CHEMICALS, INCORPORATED, 7201 HAMILTON BOULEVARD, ALLENTOWN, PENNSYLVANIA 18195, USA

[2]TECHNICAL SERVICE MANAGER – SPECIALITY ADDITIVES, AIR PRODUCTS NEDERLAND BV, KANAALWEG 15, PO BOX 3193, NL-3502GD UTRECHT, THE NETHERLANDS

ABSTRACT

Airless spray applied industrial maintenance topcoats can suffer from defects such as microfoam, gloss loss, retraction and a reduction in corrosion resistance. By careful selection of additives, such as dispersants, defoamers and wetting agents, the coatings chemist can reduce or eliminate these defects. Basic and empirical data are presented that will demonstrate how these defects were managed in several model industrial maintenance coating formulations.

INTRODUCTION

The spray application of coatings satisfies the requirement for ease of application, however it produces its own unique set of problems for the formulations chemist. This high speed, high shear application technique can cause air to be entrained in the coating which, during air drying, will rise to the surface and cause surface irregularities. These irregularities are called in the industry, external microfoam, and will cause loss of gloss. Also, entrained air that has not risen to the applied coating surface will be trapped in the final dried coating. This internal microfoam as it is referred to in the industry, can cause water sensitivity and loss of corrosion resistance.

Additionally, the airless spray application of a coating coupled with incomplete substrate preparation places a stringent requirement upon the surfactant system. The surfactant package, if not designed properly, will not prevent dewetting, and coating esthetics and substrate protection will not be maximised.

This study utilised an additive package, based on a dispersant, defoamer and a wetting agent, in a concerted approach to minimise external/internal microfoam and surface dewetting and maximise gloss. The effect of the successful additives on persistency, corrosion resistance, storage stability and pigment grind characteristics were also studied. Possible mechanisms of microfoam formation and its elimination by additives are given. Model thermoplastic water-borne industrial maintenance coating formulations were chosen that are typically airless spray applied to rail cars, exterior storage tanks and bridges.

CONCEPTS / THEORY

Airless spray is a hydraulic technique that forces paint to be atomised through a small orifice at high pressure, typically 2500 psi. With working pressures of 2500 psi, a high volume of paint can efficiently be applied over a large area. Conventional spray technique operates typically at 10-40 psi fluid pressures, and air is intentionally injected into the fluid at 30-85 psi to assist in atomisation[1]. The microfoam problem occurs actually in both types of spray operations, but is typically more severe in airless spray.

The variables affecting air entrainment in a spray applied coating are numerous. Beginning with the spray process[2,3,4], the high shear found at the spray nozzle can certainly influence the degree of air entrainment. The size of the atomised droplet may also influence the degree of air entrainment. The question is just how does the degree of air entrainment in a droplet track droplet size? As droplets get smaller, will they support less entrained air or will the ratio of entrained air to droplet volume remain the same? The latter can be the case if the entrained air bubbles get smaller as droplet size decreases. The former can be the case if air bubbles tend to form at a fixed size and are therefore reduced in number in smaller droplets. Since the size of the atomised droplet may influence the degree of air entrainment, it is pertinent to address the factors affecting atomisation efficiency.

Factors reducing average droplet size during atomisation are:

1) increased fluid pressure
2) decreased fluid flow rate
3) decreased orifice size in spray nozzle
4) reduced fluid density
5) decreased fluid viscosity at high shear
6) reduced surface tension of the fluid

The factors influencing average droplet size or atomisation efficiency must be optimised, but within the constraints of all the required performance/application specifications of the coating. For example, higher fluid rates and a larger orifice size will allow for spray of more coating per unit time but at the cost of fineness of finish due to larger spray droplets. Finer finishes can be obtained with smaller droplets by increasing fluid pressure and/or decreasing orifice size and fluid rate.

Reduction of coating density by the addition of water or solvent will reduce droplet size, but at the expense of increasing drying time and the volatile organic compound level (VOC) of the coating, respectively. In the attempt to achieve as high a solids level as possible to minimise dry time/VOC's and yet meet the required performance/spray specifications, coating density is one variable that the coating formulator needs to optimise. Additionally, a high coating solids level will affect the high shear viscosity[5,6]. An increased viscosity at high shear will reduce atomisation efficiency. Eliminating thickeners, which can contribute to an elevated high shear viscosity would be an appropriate measure. Thickeners can be chosen that will predominantly only influence low shear application properties.

A reduction in equilibrium surface tension (EST) will reduce droplet size[3] and will also provide for a more robust coating by allowing for wetting even over improperly treated surfaces. It is speculated that a reduction in dynamic surface tension (DST) may also reduce droplet size[7,8]. Dynamic surface tension, the non-equilibrium value of surface tension, is lowest for surfactants that diffuse rapidly in solution. Whereas a rapidly diffusing surfactant can quickly eliminate high surface tension at a newly created interface such as in a high speed airless spray coating application, it is speculated that the atomisation process can benefit by this process also. Once having determined the effect of atomisation efficiency on air entrainment, formulating variables such as density, percent solids, viscosity, surface tension, hardware, and spray conditions need to be optimised. This is contingent upon predetermined performance specifications and cost constraints for a given formulation.

Entrained air that has been introduced into a spray applied coating can be released either during the drying phase or possibly during transit of the atomised droplet. The ability of a spray applied coating to release its entrained air is related to:

1) bulk viscosity
2) surface viscosity at the air/liquid coating interface
3) surface viscosity at the entrained air bubble/liquid coating interface
4) rate of increase of #1 - #3 during drying
5) size of the entrained air bubbles
6) variation in dry film thickness (DFT)
7) temperature and relative humidity (RH)

Air release will be hampered by too high a low shear bulk viscosity and by too high a surface viscosity (many times referred to as just surface skinning but actually more complex) at the coating/air interface. Rapid build up of these viscosities during drying contributes to the problem. Although the use of rheology modifiers is necessary to have the correct application properties, such as flow and levelling and sag resistance, they must not be overused. The use of slower evaporating solvents to minimise the viscosity build during the critical "time window of opportunity" for air release is today more restricted due to the requirement of lower VOC's. The use of additives can keep these viscosities at a minimum.

For instance, the use of effective dispersing agents will maintain low bulk viscosity at higher pigment solids loading levels and provide adequate particle size reduction during the grind.

The use of select surface active agents can also minimise surface viscosity. The concept of surface viscosity as a mechanism for the stabilisation of foam is well documented[9,10,11]. The foam stability of shaving cream, fire fighting foam and Meringue is testament to this powerful foam stabilising mechanism. In *Figure 1*, a conventional surfactant, such as an alkylphenol ethoxylate and an acetylenic diol, are shown adsorbed within the liquid lamella of foam. The orientation of the conventional surfactant, normal to the interface, allows for the intermolecular association of adsorbed surfactant. This structure development due to surfactant cohesive strength causes an increase in surface viscosity and will stabilise foam by inhibiting drainage and can be a contributor towards surface skinning. Acetylenic diols, however, have a central hydrophilic group and are branched[12] and orient parallel to the interface. It is speculated that structure development and, therefore, also foam and surface skinning are minimised. By analogy, the adsorption of additives at the air/liquid interface of a rising bubble (entrained air) should not contribute to a high surface viscosity at this interface, or the resulting drag will also minimise the ultimate air escape from the drying coating.

The retardation of a rising air bubble by Marangoni flow[13] is yet another mechanism to be considered (*Figure 2*). A rising bubble is not rigid and it has been shown that it has an aerodynamic advantage to a rigid sphere[14]. In *Figure 2*, surfactant is shown minimising this advantage. As coating flows past the bubble, the bubble is elongated and this provides an aerodynamic advantage to the rate of rise. In this scenario, adsorption of the surfactant at the air/liquid interface occurs preferentially on the downstream portion of the bubble compared to the upstream portion. This concentration gradient, resulting in an interfacial tension gradient, will cause a liquid interfacial flow (Marangoni flow) from the surfactant rich downstream region to the surfactant poor upstream region. This Marangoni flow will reduce the aerodynamic advantage originally set up by the flow of coating past the bubble. If the advantage is completely lost, the bubble rate of rise is that of a rigid sphere given by Stokes law.

$$v = 2a^2(p_2 - p_1) \, g/9\eta$$

where: v = air bubble rate of rise
　　　　a = radius of the rising air bubble
　　　　p_2 = liquid medium density
　　　　p_1 = air bubble density
　　　　g = gravitational constant
　　　　η = liquid medium viscosity

Surfactants that diffuse rapidly may minimise this Marangoni flow by minimising the surface tension gradients, thereby allowing for a more rapid rate of rise of entrained air in an applied coating. One technique used to compare surfactant diffusion of one surfactant class to another is the measurement of dynamic surface tension.

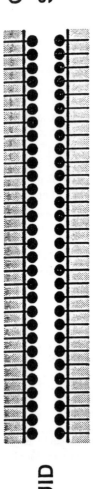

Figure 1 – Mechanism of Foam Stabilization – Surface Viscosity

A conventional surfactant is shown vertically adsorbed and an acetylenic diol horizontally adsorbed in the lamella of foam. Intermolecular attraction of the oriented conventional surfactant is responsible for an increase in surface viscosity whereas the reduction of intermolecular attraction between oriented acetylenic diol molecules minimizes surface viscosity and therefore foam stability.

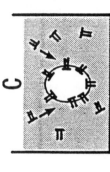

CONVENTIONAL SURFACTANT

A   B

LIQUID COATING

AIR BUBBLE

SURFYNOL SURFACTANT

C

Figure 2 - Mechanism for Release of Entrained Air.

A rising air bubble in (A), with adsorbed and oriented surfactant at its interface, will experience a surfactant gradient and, therefore, also a surface tension gradient at the air/liquid interface. Surface transport of surfactant will cause a Marangoni flow in the direction of the arrows. Result: The rising bubble is forced into a spherical shape, as in (B), and loses its aerodynamic geometry. Bubble rate of rise is minimized.

In (C), bulk transport of surfactant to the surfactant-deficient region eliminates the surface tension gradient at the air/liquid interface. Result: No Marangoni flow occurs and the rising air bubble maintains an aerodynamic geometry to maximize rate of bubble rise.

Table 1 identifies comparative EST/DST surfactant performance in deionized water as measured using the maximum bubble pressure technique[15]. 2,4,7,9-tetramethyl-5-decyne-4,7-diol (TMDD) is shown to be a rapidly diffusing surfactant as evidenced by the low dynamic surface tension.

The magnitude of the buoyancy force causing the air bubble to rise is also dependent on the size of the air bubble. Larger bubbles in an applied coating will more rapidly rise to the surface and break than smaller bubbles. To maximise the size of each air bubble, only those surfactants that do not hinder bubble collision and coalescence during the spray processes should be employed. From *Figure 3*, it can be seen that surfactants such as alkylphenol ethoxylates with long hydrophilic groups can entropically hinder effective air bubble collisions that would result in the growth of larger bubbles. Also, a high surface viscosity as discussed above will have a similar opposing effect on large bubble growth.

Table I

SURFACE TENSION COMPARISONS

MAXIMUM BUBBLE PRESSURE TECHNIQUE

	SURFACE TENSION (dynes/cm) (.1% Concentration)		
	Equilibrium S.T. (1 Bubble/Sec.)	Dynamic S.T. (6 Bubble/Sec.)	ΔST
Linear C12-15 Alchohol Ethoxylate	32.5	55.8	23.3
Branched C13 Ethoxylate	31.5	44.7	13.2
Octyphenol + 10EO	33.4	44.6	11.2
Sodium Lauryl Sulfate	43.8	53.0	9.2
Dioctyl Sodium Sulfosuccinate	31.7	33.6	1.9
Tetramethyl Decynediol	32.5	36.5	4.0
Tetramethyl Decynediol + 1.3 Moles Ethylene Oxide	32.0	35.1	3.1
Tetramethyl Decynediol + 3.5 Moles Ethylene Oxide	34.0	37.3	3.3

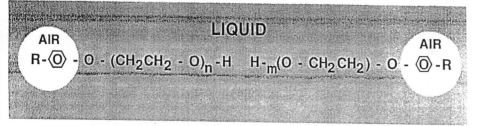

Figure 3 - Mechanism of Foam Stabilization - Steric Hindrance
The coalescence of dispersed foam bubbles in a liquid media into large unstable bubbles is retarded due to the pendant hydrophilic groups preventing the close approach of the two lamella.

As the DFT of an applied coating increases, the degree of air entrainment remaining in the dry coating and the severity of the surface defects caused by escaping air frequently increases. Also since the rate of evaporation of an applied air dried coating is related to temperature/RH it is readily seen that the extreme condition of high temperature and low RH will minimise the "time window" for escape of entrained air.

Given the complexity associated with the variable affecting atomisation droplet size and air entrainment stability, this study focused on the use of inherently low foam additives. Additionally, some of these additives exhibit superior EST/DST reduction properties. For those additives or combinations that reduced microfoam, increased gloss and had a beneficial effect on wetting unclean surfaces, their continued long term effect in the coating system, effect on air entrainment suppression, high/low shear viscosity, atomised droplet size and corrosion resistance were studied. Temperature and relative humidity were rigidly controlled during airless spray so as to keep constant their effect on the "time window".

EXPERIMENTAL DETAILS

Appendix I lists all the chemical raw materials/manufacturers and Appendix II lists all the equipment used in the study.

Coatings Preparation

Four model industrial maintenance coatings supplied by the resin manufacturers were studied. Maincote® HG-54 and Maincote AE-58 (one component water-borne acrylic and two component acrylic-epoxy, respectively, Rohm and Haas, Inc.), Aquamac™ 700 (one component acrylic, McWorther, Division of Valspar), and SCX™-1520 (one component acrylic, S.C. Johnson Polymers) were the resin types employed in the model formulations. (See Formulations A, B, C and D for general formulation details supplied by the latex manufacturer).

A Premier Mill was used for preparing the grind. Grind efficiency and viscosity were measured using the Hegman NS scale on a fineness of grind gauge and a Brookfield Viscometer, respectively. Final coating viscosity, in Krebs Units, was adjusted for airless spray per the latex manufacturers recommendation by adding water.

Coating Evaluation

All coatings were aged 24 hours before all evaluations, then Stormer viscosity was checked for drift. Coatings were prepared such that after 24 hours, viscosity was always in specification.

Viscosity at low (0.1 sec.$^{-1}$) and high (10,000 sec.$^{-1}$) shear was measured with a Rheometrics Dynamic Analyser, RDA II, and an ICI Viscometer, respectively.

The general air release property of the coating was measured by high shear mixing with a Waring Blender for 60 seconds at 12,000 rpm followed immediately by determining coating density.

All drawdowns were performed using a Wirecator to deliver 3 mils dry film thickness on metal test panels. Drying rates and gloss were measured on the drawdowns using a Paul N. Gardner Bi-Cycle Drying Time Recorder and a Hunderlab Glossmeter, respectively.

All coatings were airless sprayed at 71-73 degrees F and 61-65% relative humidity. To eliminate human error, all spray was performed by an automatic spray machine with a 5 inch index distance and 1500-1600 inch/minute traverse speed so as to achieve a 3.5 +/- 0.5 mil. dry film thickness (DFT) on all metal test panels.

Average particle size of the spray droplets after atomisation was determined by spraying black test paper with the white coatings by allowing only a small fraction of the spray to reach the paper, thereby maximising the separation of droplets on the test paper. This technique was accomplished by providing for a 1/8 inch opening just in front of the test paper and in the centre of the spray pattern and dropping a shutter so as to minimise the amount of spray deposited on the paper.

A microscope and image analyser were used for droplet visualisation and ultimate particle size measurement. The field of vision was chosen for the microscope such that approximately 1000 droplets could be analysed per view. Eight to nine views were made so that the total number (approximately 8000) of droplets being analysed was high enough to allow for the statistical treatment of the data.

External and internal microfoam was assessed subjectively by visual means and profilometry was used to quantify external microfoam through changes in average surface roughness and to serve as a check on the visual assessment technique.

Salt spray evaluations were performed according to ASTM B117, D-610, D-714 and D-1654 to determine the effect of additives on corrosion resistance. All coatings were sprayed direct to metal on sandblasted, 3 mil profile, hot rolled steel.

Formulation A
HIGH-BUILD, GLOSS, WHITE TOPCOAT -- MAINCOTE HG-54

GRIND PREPARATION

Grind in Cowles Dissolver 15 mins. then let down at slower speed

Formula	Control (HB-54-1)	A	B	C
		POUNDS		
INGREDIENTS				
Maincote HG-54	400.0	400.0	400.0	400.0
Pigment Dispersant				
Tamol 681	9.1	-	-	9.1
Surfynol CT-151	-	**8.5**	**8.5**	-
NH$_4$OH (28% NH$_3$)	1.0	1.0	1.0	1.0
Grind Defoamer				
Drew L-405	2.4	-	-	-
Surfynol DF-210	-	**2.4**	**2.4**	**2.4**
Ti-Pure R-900 HG	127.6	127.6	127.6	127.6
Allow pigment to wet thoroughly, then add:				
Rheology Modifier QR-708	0.5	0.5	0.5	0.5
LETDOWN PREPARATION				
Maincote HG-54	318.5	318.5	318.5	318.5
Texanol	44.7	44.7	44.7	44.7
Letdown Defoamer				
Drew Y-250	3.5	-	-	-
Surfynol DF-210	-	**2.4**	**2.4**	**2.4**
NH$_4$OH (28% NH$_3$)	4.0	4.0	4.0	4.0
Sodium Nitrite (15% Aq. Solution)	8.2	8.2	8.2	8.2
Water	29.0	29.0	29.0	29.0
Rheology Modifier QR-708	0.5	0.5	0.5	0.5
Wetting Agent				
Surfynol 104DPM	-	19.0	-	19.0
Surfynol 61	-	-	**9.5**	-
Total	949	966.3	956.8	966.9

Formulation B
GLOSS, WHITE, ACRYLIC/EPOXY TOPCOAT
MAINCOTE AE-58

Acrylic Component A

GRIND PREPARATION

Grind the following materials using a high speed dissolver for 20 minutes:

Formula	Control (G-58-1)	A
INGREDIENTS	POUNDS	
Methyl Carbitol	38.8	38.8
Tamol 165	13.8	—
Surfynol CT-151	--	8.4
NH_4OH (28% NH_3)	1.0	1.0
Triton CF-10	1.6	—
Surfynol 104DPM	--	3.2
Patcote 519	0.4	—
Surfynol DF-60	--	0.6
TiPure R-900	193.7	193.7

Add the following and continue to grind for 2-3 minutes at lower speed:

Water	19.9	19.9
Total grind	269.2	265.6

LETDOWN PREPARATION

Add the following in the order listed and mix thoroughly:

Maincote AE-58	493.0	493.0
Water	58.5	58.5
NH_4OH (28% NH_3)	2.4	2.4
Grind (from above)	269.2	265.6
Ektasolve EEH	48.2	48.2
Patcote 531	2.0	—
Surfynol 104DPM	--	16.8
Water	14.2	14.2
Acrysol RM-1020	8.0	—
QR-708	1.2	1.2
Sodium Nitrite (15% aqueous solution)	8.8	8.8
Total Acrylic Component A	905.5	908.7
Surfynol DF-60	--	6.5

Epoxy Component B

Genepoxy 370-H55	94.8	94.8
TOTAL ACRYLIC/EPOXY TOPCOAT	1000.3	1010.0

Formulation C

HIGH GLOSS WHITE
AQUAMAC 700
700 st1

GRIND PREPARATION
Disperse in high speed dispersator to NS 7 grind.
Add to latex under agitation when grind is achieved.

INGREDIENTS	POUNDS
Tap Water	71.5
Tamol 681	6.6
Surfynol 104A	**2.0**
Rheology Modifier RM-825	1.0
Bubble Breaker 3056A	1.0
RCL-535-TiO$_2$	210.0

LETDOWN

Aquamac 700	574.2
Tap Water	65.3

Add to adjust pH between 8 and 9

Aqueous Ammonia (28%)	2.0

Add the following

Texanol	71.1
Sodium Nitrite (4%)	9.0
	1013.7

Formulation D

HIGH-GLOSS WHITE
SCX-1520
92-196-U

GRIND PREPARATION
Disperse at High Speed to 5-7 HEGMAN

INGREDIENTS	POUNDS
Joncryl 56	126.2
Ti-Pure R-900	141.6
Byk 020	1.93
Water	19.9

LETDOWN

Water	33.8
SCX-1520	482.2
Dowanol EB	43.5
Dowanol DB	14.5
10% Ammonium Benzoate in water	29.1

PREMIX THE FOLLOWING:

Surfynol 104H	**4.82**
Byk 020	1.45

THEN ADD AS REQUIRED

Water	56.8
	955.8

RESULTS / DISCUSSION

Defect Control - An Additives Solution

In this study, microfoam was observed as surface irregularities in the dry applied coating and were approximately 0.1-0.2 mm in diameter. Airless spray application accounts for the microfoam since the dry coating, applied via drawdown technique, microfoam is not present. *Figure 4* shows examples of both external and internal microfoam. It is presumed that the high shear/high pressure to which the coating is subjected is responsible for the problem. Possibly, air is dissolved in the coating and is partially released during transit of the atomised droplet to the substrate and during drying.

Three additive packages have been identified that will minimise microfoam, maximise gloss, promote substrate wetting on improperly cleaned substrates, not diminish corrosion resistance and be persistent in airless spray applied water-borne industrial maintenance topcoats. An additive systems approach involving dispersant, defoamer and wetting agent was taken to achieve our objectives. At the outset, select additives and/or surfactants were chosen for investigation that:

1) are inherently low foaming
2) confer both low EST/DST
3) minimise surface viscosity due to low coherence at an interface
4) do not entropically hinder entrained air bubble coalescence/release

Maincote HG-54 Based Model Formulation-Results

Figure 4 shows the effect of airless spraying on external microfoam formation in a control Maincote HG-54 emulsion based formulation (Formulation A) compared to one additive package (Formula A in Formulation A) that will reduce internal/external microfoam by up to 90%.

Table 2 shows in detail the performance for all three additive systems. Results show that the recommended starting formulation for Maincote HG-54 will lose gloss after airless spray compared to the gloss obtained on drawdown. The three additive systems, A, B and C, based on a select dispersant, defoamer and wetting agent combination, provide much resistance to microfoam and gloss loss upon spray application. Subjective visual assessment of external/internal microfoam tracked gloss performance. Depending upon the additive system chosen, as much as 90% of the microfoam can be eliminated. For verification of the subjective assessment of external microfoam, average surface roughness was measured by profilometry and the trends agreed well. *Figure 5* shows the profilometry profiles for airless spray applied Formula A and the control HB-54-1.

EXTERNAL MICROFOAM

Control Formulation – HB-54-1 (Formulation A)

Drawdown   Airless Spray

Airless Spray

Additive Package (Formula 1, Table 1):
Dispersant: Surfÿnol CT-151 • Defoamer: Surfÿnol DF-210 • Wetting Agent: Surfÿnol 104DPM

INTERNAL MICROFOAM

Airless Spray

Control HB-54-1   Additive Package
                                                          (Formula 1, Table 1)

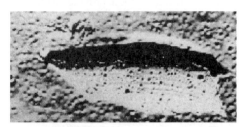

Figure 4 - Maximum Reduction of Microfoam in Airless Spray
Photomicrographs - Magnification 15x
Resin: Maincote HG-54

Table 2

EFFECT OF ADDITIVES ON GLOSS/MICROFOAM

Initial Test: 24 Hours After Coating Preparation

Resin: Maincote HG-54

FORMULA	Control (HB-54-1)	ADDITIVE SYSTEM		
		A	B	C
Dispersant	Tamol 681	Surfynol CT-151	Surfynol CT-151	Tamol 681
Defoamer	Drew L-405/Y-250	Surfynol DF-210	Surfynol DF-210	Surfynol DF-210
Wetting Agent	No Recommendation	Surfynol 104DPM	Surfynol 61	Surfynol 104DPM
PERFORMANCE PROPERTIES				
Gloss				
Drawdown				
20°	33.7	33.7	33.6	29.5
60°	78.1	77.3	73.4	74.8
Airless Spray				
20°	13.3	31.0	32.3	26.1
60°	54.0	71.7	72.6	70.2
% Microfoam Reduction				
Visual				
External	0	75-90	75-90	75
Internal	0	75-90	75	75
From Average Surface Roughness[1]				
External	0	74	73	63

[1]Tencor P-2 Long Scan Profiler

Resin: Maincote HG-54 (Control—HB-54-1)

PROFILOMETRY CURVES RESIN SYSTEM: MAINCOTE HG-54 CONTROL—HB-54-1

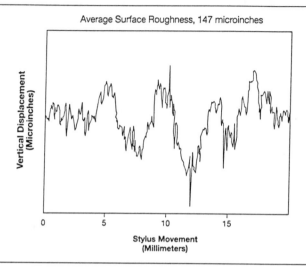

SURFȲNOL ADDITIVE PACKAGE—FORMULA 1, TABLE 1

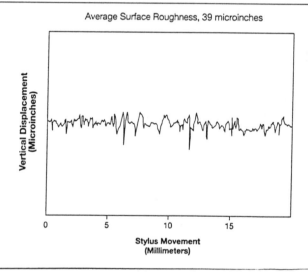

Figure 5 - Profilometry Curves

The additive packages in *Table 2* are based on the following chemistries:

SURFYNOL® CT-151 anionic dispersant	ammonia neutralised polycarboxylic acid
SURFYNOL® 104DPM surfactant	50% solution of 2,4,7,9-tetramethyl-5-decyne-4,7-diol in dipropylene glycol monomethyl ether
SURFYNOL® 61 surfactant	3,5-dimethyl-1-hexyn-3-ol
SURFYNOL® DF-210 defoamer	non silicone organic defoamer

To insure that additive performance longevity is maintained, the data in *Table 3* shows that Formula A will maintain effectiveness in accelerated storage testing at both 120 F and 140 F for 2 weeks and for 10 days respectively. Although not tested, it is expected that Formulas B and C would perform similarly.

Care must be exercised in the choice of additives since both anionic dispersing agents and surface active agents can reduce the corrosion resistance of the coating. *Table 4* shows the results of corrosion resistance testing for Formulas A, B and C. All three additive packages were shown not to cause loss of corrosion resistance. Actually a mild improvement was observed. It needs to be noted that the corrosion testing was performed according to ASTM standards and that even though the formulations are designed as topcoats, they were applied directly to sandblasted, hot rolled steel with a 3 mil. profile. In actual practice, the topcoat would be applied over a primer, but this technique would mask most of the additive effect on corrosion resistance.

Although the above results show that tetramethyl decynediol does not adversely affect corrosion resistance, a formulator can take an alternative approach and employ 3,5-dimethyl-1-hexyn-3-ol, a volatile surfactant, which will evaporate from the air drying film and therefore leave no possible residue to cause either water sensitivity or corrosion resistance problems (Formulation A, Formula B gives the complete formula description and *Table 2*, Formula B gives the performance data).

As part of the recommended additive package to reduce microfoam, SURFYNOL CT-151 dispersant, a low foam type anionic dispersant, also maintained effectiveness in accelerated storage testing. During accelerated storage testing described above, coating containing SURFYNOL CT-151 dispersant maintained a stable dispersion and comparable viscosity when compared to the control formulation. Additionally, SURFYNOL CT-151 dispersant improved the grind by more efficiently reducing the particle size with no adverse effect on grind viscosity compared to the polyacrylate dispersant based coating (*Figure 6*).

Table 3

EFFECT OF ADDITIVES ON GLOSS/MICROFOAM ON AGING
Longevity Test: 2 Weeks @ 120° F
Resin: Maincote HG-54

FORMULA	ADDITIVE SYSTEM	
	Control (HB-54-1)	1
Dispersant	Tamol 681	Surfynol CT-151
Defoamer	Drew L–405/Y-250	Surfynol DF-210
Wetting Agent	No Recommendation	Surfynol 104DPM

PERFORMANCE PROPERTIES

Gloss
  Drawdown

20°	28.0	28.5
60°	70.8	68.0

  Airless Spray

20°	21.0	26.7
60°	62.8	69.6

% Microfoam Reduction

  Visual

External	0	50-75
Internal	0	50-75

Table 4

CORROSION TEST RESULTS (ASTM B117)

SALT SPRAY: 250 HOURS

Resin: Maincote HG-54

FORMULA	Control (HB-54-1)	ADDITIVE SYSTEM		
		A	B	C
Dispersant	Tamol 681	Surfynol CT-151	Surfynol CT-151	Tamol 681
Defoamer	Drew L-405 / Y-250	Surfynol DF-210	Surfynol DF-210	Surfynol DF-210
Wetting Agent	No Recommendation	Surfynol 104DPM	Surfynol 61	Surfynol 104DPM

CORROSION TEST RESULTS

	Control	A	B	C
Rust (ASTM D-610)	4	4	5	5
Blisters (ASTM D-714)	2M	6M	4F	4M
Scribe (ASTM D-1654)	6	7	7	7

Abbreviated Key:

RUST
0–Approximately 100% of surface rusted
5–Rusting to the extent of 3% of surface rusted
10–no rusting or less than 0.01% of surface rusted

BLISTERS
0–10 Blister size is inversely proportional to #
M–Medium
F–Few

SCRIBE
0–Larger than 16mm
5–Between 3 and 5mm
10–No creepage from scribe

(A) Particle Size Reduction

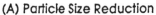
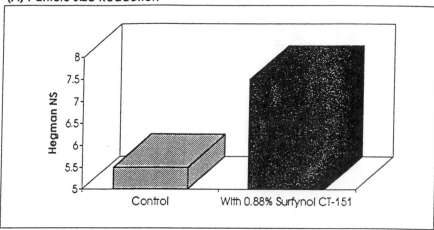

(B) Grind Viscosity

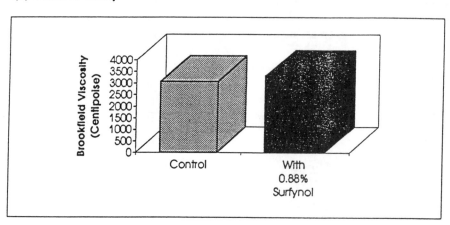

Figure 6   Comparative Grind Dispersion/Viscosity Characteristics
            Surfynol CT-151 dispersant is more efficient than the control
            dispersant, with no adverse effect on grind viscosity
            (Resin System: Maincote HG-54)

Maincote AE-58 Based Model Formulation-Results

Similar to the results with Maincote HG-54, *Table 5* shows the performance of one
additive package (for additive description see Appendix I) based on a select dispersant,
defoamer and wetting agent in a Maincote AE-58 emulsion based starting formulation
(Formulation B). Results show that the highest gloss will result after airless spray for the
new recommended additive package. The lower gloss on draw down for the new package
compared to the control coating is unexplained. Possibly improved mixing of the
additives occurred upon spray. It needs to be noted that although microfoam was reduced
by 33%, in the Maincote AE-58 system, the absolute severity of the problem at 3-4 mils
was much lower than with the Maincote HG-54 emulsion based system (See *Table 6* for
comparative average surface roughness values). In this case, improvement in gloss is
probably also due to improved dispersion characteristics and/or compatibility of select
components. For instance, if pigment flocculation occurs at the surface of the drying
coating due to incompatible coating components rising to the surface or due to simply
a poor choice of dispersant, the gloss will be reduced. It is speculated that the
recommended additive package probably improves gloss due to some combination of the
aforementioned possible mechanisms.

Aquamac 700 Based Model Formulation-Results

The starting model formulation supplied by McWhorter recommends the use of
SURFYNOL 104A surfactant in the grind (Formulation C). *Figure 7* shows a loss of
grind efficiency (a half unit on the Hegman grind gauge) and a large increase in grind
viscosity for a 15 minute grind, if the SURFYNOL 104A surfactant is left out of the
grind. SURFYNOL 104A surfactant is low foaming, hydrophobic, contributes little to
water sensitivity and does not detract from corrosion resistance, and is clearly functioning
as a grind aid. Presumably, it preconditions the surface of the pigment and thereby
maximises the viscosity reduction effectiveness of the dispersant.

All Model Formulations-Wetting Properties

2,4,7,9-tetramethyl-5-decyne-4,7-diol was shown to be an exceptional wetting agent
providing much control over dewetting. In all four industrial maintenance coating
formulations containing additive packages, dewetting was controlled in either airless
spray or drawdowns on lubricant contaminated steel test panels (*Figure 8*). Depending on
the level of contaminant, either the complete elimination or a significant reduction of
retraction is achieved. 2,4,7,9-tetramethyl-5-decyne-4,7-diol, previously shown to not
adversely affect corrosion resistance in a Maincote HG-54 based coating, will provide for
a more robust coating. As a primary measure against dewetting, it is recommended that
substrate cleaning and preparation be as complete as possible and that surfactant be relied
upon only as a secondary measure of protection.

Table 5

EFFECT OF ADDITIVES ON GLOSS/MICROFOAM
Initial Test: 24 Hours After Coating Preparation
Resin: Maincote AE-58

FORMULA	ADDITIVE SYSTEM	
	Control (G-58-1)	A
Dispersant	Tamol 165	Surfynol CT-151
Defoamer	Patcote 519/ Patcote 531	Surfynol DF-60
Wetting Agent	Triton CF-10	Surfynol 104DPM
PERFORMANCE PROPERTIES		
Gloss		
Drawdown		
20°	37.1	20.2
60°	78.5	61.8
Airless Spray		
20°	24.6	30.9
60°	71.4	75.1
% Microfoam Reduction		
Visual		
External	0	25-50
Internal	0	25-50
From Ra[1], Average Surface Roughness		
External	0	33

[1] Tencor P-2 Long Scan Profiler

Table 6

Average Surface Roughness Of Airless Spray Applied Industrial Coatings

Comparison Of The Starting Point Control Additive Package To The New Additive Package

Resin System:	Maincote HG-54 (See Formulation A)		Maincote AE-58 (See Formulation B)	
Additive Package	Control	Formula A (New Additive Package).	Control	Formula A (New Additive Package).
Dispersant	Tamol 681	SURFYNOL CT-151	Tamol 165	SURFYNOL CT-151
Defoamer	Drew L405/Y250	SURFYNOL DF-210	Patecote 519/531	SURFYNOL DF-60
Wetting Agent	None	SURFYNOL 104DPM	Triton CF-10	SURFYNOL 104DPM
Average Surface Roughness				
Ra, microinches	147	39	21	15

(A) Particle Size Reduction

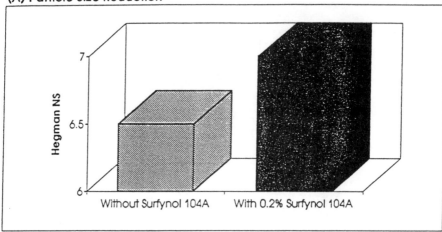

(B) Grind Viscosity

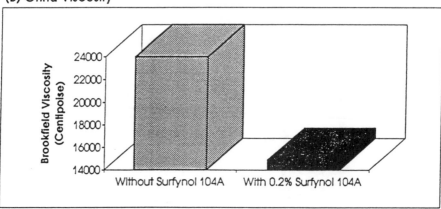

Figure 7   Comparative Grind Dispersion/Viscosity Characteristics
              Surfynol 104A grind aid improves the grind efficiency
              and assists the dispersant in grind viscosity reduction
              (Resin System: Aquamac 700)

Resin: Joncryl SCX-1520 (Formulation D)
Application Technique: Drawdown

No Surfÿnol 104H

With 0.5% Surfÿnol 104H

Resin: Aquamac 700 (Formulation C)
Application Technique: Drawdown

o Surfÿnol 104A

With 0.6% Surfÿnol 104A

Resin: Maincote AE-58 (Formulation B)
Application Technique: Airless Spray

o Surfÿnol 104DPM

With 2.0% Surfÿnol 104DPM

Resin: Maincote HG-54 (Formulation A)
Application Technique: Airless Spray

o Surfÿnol 104DPM

With 2.0% Surfÿnol 104DPM

gure 8 - The Effect of Surfÿnol Surfactants on Substrate Wetting. Test Panels Contaminated with Lubricant Oil

A summary chart (*Table 8*) is provided that lists the resin systems studied along with defects that select SURFYNOL surfactants, at specific recommended use levels, can minimise or eliminate.

It is emphasised that the variables that were held constant for the study were the temperature and RH during spray and air dry, at 71-73 degrees F, and 61-65%, respectively. DFT's were kept at 3+/-0.5 mils. The assumption/desire is that at mildly divergent conditions from the above, the additive package(s) would still prove useful. The concept of coatings performing under a broad window of use conditions is not new. Additives that provide for robust performance are necessary to the commercial success of the coating. Probably at severely divergent conditions from the above, no additive package alone could control microfoam.

Microfoam Formation/Elimination-Possible Mechanism

The Maincote HG-54 topcoat, based on the control and one new additive package (Formulation A; Control HB-54-1 and Formula A), was studied further to reveal information on the mechanism of microfoam formation and elimination. For these coating systems, the following properties were determined:

> average atomised droplet size
> general air release properties of the coating in a high shear environment
> high and low shear viscosity

Again, looking first at the atomisation process, we found a small reduction of microfoam, 10-25%, in airless spray application of the control formulation resulted if the:

> spray nozzle orifice size is reduced
> fluid delivery rate is reduced
> liquid coating spray pressure is increased

This suggests that for this formulation, improved atomisation efficiency (lower droplet size) will influence the degree of air entrainment. We compared the "average droplet size" for both the control and the new additive based coating and found the control formulation's "average droplet size" to be 36% larger (*Table 7*).

The technique used here (recommended by Kwok and Liu[3]) to measure "average droplet size" (see Experimental Details Section) briefly consisted of allowing a small fraction of the center of the spray pattern to be deposited on black paper followed by measurement with microscopic/image analysis techniques. Ideally, laser techniques[3,4] would actually measure the average droplet size in a very large volume of spray droplets and would be more diagnostic. Nonetheless, these results are encouraging and suggest that air that is incorporated into this coating during the spray process has a "standard size" and a smaller droplet, will therefore result in a lower air to liquid ratio.

Table **7**

Performance Properties of the Recommended Starting Formulation Compared To The New Additives Based System

Resin System: Maincote HG-54
(Formulation A)

	Control Formulation	New Additives Package
Additive Package		
Dispersant	Tamol 681	SURFYNOL CT-151
Defoamer (s)	Drew L-405/Y-250	SURFYNOL DF-210
Wetting Agent	-	SURFYNOL 104 DPM
Properties		
% Microfoam Reduction	0	75-90
Wt. % Solids		
Drawdown-Experimental	49.5	50.8
After Spray-Experimental	49.8	51.0
Increase Due To Spray	0.3	0.2
Calculated-Resin +Pigmt	44.9	44.1
"Average Particle Size",		
μm +/- 3σ	986 +/-54	724 +/-43
High Shear Foam Test	0.86	0.93
Calculated-Resin +Pigment		
Solids, Wt. %	46.4	45.6
Density, g/ml,		
Unagitated	1.14	1.14
Agitated	0.86	0.93
Agitated, No Defoamer	0.86	-
Viscosity		
Krebbs Units, KU	99 +/- 1	99 +/1
ICI, @ 10,000 sec-1, Poise	0.41	0.41
DMS @ 0.1 rad/sec, Poise	1126	819
Dry Time, Minutes	30-35	25-30

Table 8

SUMMARY CHART

Resin	Problem Addressed	Product Type	Surfynol Recommendation	Percentage on Final Coating	
				Grind	Letdown
Maincote HG-54 (Formulation A)	o External/Internal Microfoam Reduction o Low Gloss	Dispersant Defoamer Wetting Agent	Surfynol CT-151 Surfynol DF-210 Surfynol 104DPM or Surfynol 61	0.88 0.24 - - - -	- - 0.24 2.00 1.00
	o Retraction/Orange Peel	Wetting Agent	Surfynol 104DPM	- -	2.00
Maincote AE-58 (Formulation B)	o External/Internal Microfoam Reduction o Low Gloss	Dispersant Defoamer Wetting Agent	Surfynol CT-151 Surfynol DF-60 Surfynol 104DPM	0.83 0.06 0.32	- - 0.64 1.66
	o Retraction/Orange Peel	Wetting Agent	Surfynol 104DPM	- -	2.00
Aquamac 700 (Formulation C)	o Dispersancy	Grind Aid	Surfynol 104A	0.20	- -
	o Retraction/Orange Peel	Wetting Agent	Surfynol 104A	0.20	0.4
SCX-1520 (Formulation D)	o Retraction/Orange Peel	Wetting Agent	Surfynol 104H	- -	0.5

Surfactants such as TMDD, that provide for both a low EST and DST, (*Table 1*) may be responsible for the improvement in atomisation efficiency, thereby lessening the inherent air entrainment. The EST/DST of the actual coating could not be reproducibly determined either by the Du Nouy ring[16], or maximum bubble pressure technique unless the coatings were extensively diluted with water. Presumably, this was due to the coating viscosity and/or the nature of the coating/air interface.

A simple high shear mix test can reveal much about a coating's propensity to release its entrained air. Air is released in this type of test as a result of a more favourable rate of bubble rise due to:

1) Effective collisions of air bubbles during mixing, resulting in larger more buoyant bubbles.
2) Low bulk/surface viscosity.
3) Possibly, low dynamic surface tension.

The density of the coatings was determined immediately after high shear mixing. The higher the density, the less air has been retained. These density determinations were made on coatings prepared with 1.5% higher level of solids (pigment + resin) by holding out water. This was done since our interest is in the air release properties after spray and during the first two hours of drying. It was observed that most of the drying occurred over approximately two hours (*Figure 9*). A 1.5% higher solids level was arbitrary and is clearly a compromise, since it was beyond the scope of the study to continuously monitor the air release properties during two hours of drying. It would take roughly six minutes of dry time immediately after spray for the solids to increase 1.5%. Interestingly, the airless spray process only increased the solids level approximately 0.3% under the conditions of study (*Table 7*).

All additives, in the additive packages that were successful in significantly reducing microfoam, were low foaming as measured in this high shear foam test. It has been reported that polyacrylate dispersants can cause more foam in coatings than lower foaming non-polyacrylate dispersants[17]. We observed that SURFYNOL CT-151 dispersant, an anionic, non-polyacrylate pigment dispersant, is much less foamy in the grind than the resin supplier's recommended polyacrylate. Also, SURFYNOL DF-210 defoamer is more effective as a deairentraining agent than the recommended defoamer. *Table 7* shows the air release properties (density difference after high shear mixing) of the two coatings with/without the new additive package. It is possible that the new additive based coating allows for more effective bubble collisions (*Figure 3*) in transit, immediately after atomisation, and possibly also in a reduced surface viscosity, at the entrained air bubble/liquid coating interface, so that entrained air is released before the drying coating has developed too high a bulk viscosity. It should be noted that the control formulation employs no surfactant. However, in the polymer synthesis of the binder, surfactant is typically introduced which may absorb at an entrained air bubble/liquid coating interface thereby increasing surface viscosity and retarding the rate of air bubble rise. The introduction of TMDD may favourably compete with "surface viscosifiers" thereby reducing this viscosity and improving rate of bubble rise.

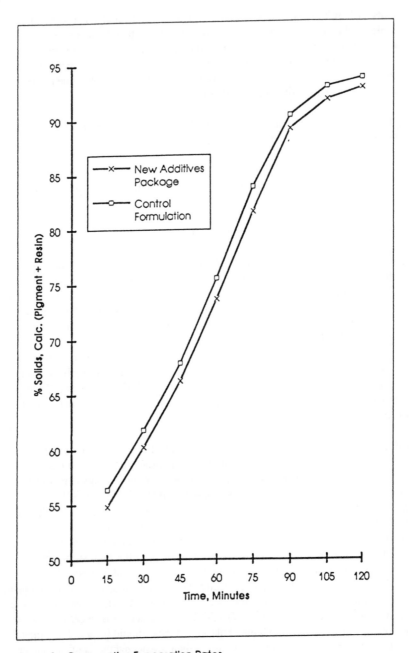

Figure 9   Comparative Evaporation Rates
The new additive package compared to the
control package, shows no appreciable difference.
(Resin System: Maincote HG-54)

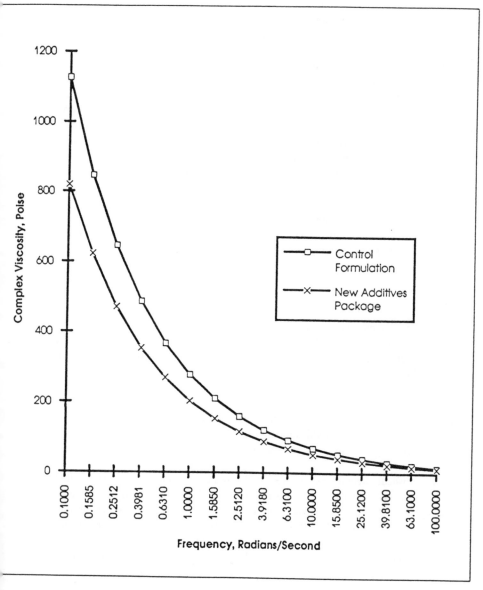

gure 10 Comparative Low Shear Viscosity
Effect of the new additive package compared to the control
package on the viscosity at low frequency (low shear),
measured with the Rheometrics Dynamic Analyzer.
(Resin System: Maincote HG-54)

Both the control and new additive based formulations were prepared with the same viscosity as per the latex supplier, namely a Stormer viscosity of 99 +/- 1 KU (roughly 1500 sec^{-1}). High shear (10,000 sec^{-1}) viscosity, measured with the ICI viscometer, was the same for both coating systems (*Table 7*). However, using dynamic mechanical spectroscopy (DMS), the complex low shear viscosity (at a frequency of 0.1-100 radians/sec) as measured with the Rheometrics Dynamic Analyser was different. From *Figure 10*, the new additives based coating had lower viscosity at low shear rates compared to the control based coating, specifically a 307 poise lower viscosity at 0.1 radians/sec. This data is also summarised in *Table 7*. This reduced low shear viscosity had no effect on sag and flow and levelling at 3-4 mils DFT and would be expected to give improved air release properties. Ideally, it would have been instructive to monitor the low shear viscosity during the first 2 hours of drying, but this too was beyond the scope of the study.

Comparative drying times and evaporation rates were performed to determine if this caused the reduction in microfoam. From *Table 7* and *Figure 10*, the drying times and evaporation rates, respectively, are not significantly different.

CONCLUSION

With an understanding of the mechanics of foam/air entrainment (de)stabilisation, several additive packages, resulting from a concerted additives systems approach, have been found to significantly reduce microfoam and maximise gloss in thermoplastic waterborne industrial topcoats under the conditions of study. The additives, consisting of a dispersant, defoamer and a wetting agent, were all found to be persistant on accelerated aging with no adverse effect on corrosion resistance. Individually, the additives contributed to improved grind efficiency, lower process or mixing foam and substrate wetting. No one additive could significantly influence the microfoam/gloss problems but, collectively, they minimised these problems.

Several physical phenomena have been addressed in an attempt to understand the mechanisms associated with microfoam formation/elimination. Of most probable consequence is the surface viscosity at the coating/air interface and at the entrained air bubble/coating interface, low shear viscosity, atomised droplet size and size of the entrained microfoam. No conclusions can be drawn, however, regarding the quantitative relative importance of each of the proposed mechanisms and this remains fertile ground for future study.

ACKNOWLEDGEMENTS

The authors wish to thank Rohm and Haas Company, S.C. Johnson Polymer and McWhorter for their support in this project by supplying latex. Additionally, M.R. Kittig, Dr. M.S. Vratsanos and J. Sheesley for average surface roughness determinations, DMS determinations and statistical mathematical support, respectively and Dr. S.W. Medina for editorial support.

Appendix I
Chemical List with Manufacturers

Ingredient	Supplier	Ingredient	Supplier
Acrysol RM-825	Rohm & Haas	Methyl Carbitol	Union Carbide
Acrysol RM-1020	Rohm & Haas	Patcote 519	Patco Ctgs Prod.
Aquamac 700	McWhorter	Patcote 531	Patco Ctgs Prod.
Bubble Breaker 3056a	Witco	RCL-535	SCM Chemicals
Byk 020	Byk Chemie	Rheology Mod. QR-708	Rohm & Haas
Cabosil M5	Cabot	SURFYNOL CT-151	Air Products
Dowanol DB	Dow	SURFYNOL DF-60	Air Products
Dowanol EB	Dow	SURFYNOL DF-210	Air Products
Drew L-405	Drew Chemical	SURFYNOL 61	Air Products
Drew Y-250	Drew Chemical	SURFYNOL 104A	Air Products
Ektasolve EEH	Eastman Chem.	SURFYNOL 104DPM	Air Products
Genepoxy 370-H55	Daubert Chem.	SURFYNOL 104H	Air Products
SCX-1520	S.C.Johnson Polymer	Tamol 165	Rohm & Haas
Joncryl 56	S.C.Johnson Polymer	Tamol 681	Rohm & Haas
Maincote AE-58	Rohm & Haas	Texanol	Eastman Chemical
Maincote HG-54	Rohm & Haas	TiPure R-900 Tipure R-900HG	DuPont
		Triton CF-10	Union Carbide

continued...

Appendix 1 (Cont'd)

Co. Address/Phone Number	Co. Address/Phone Number
Air Products 7201 Hamilton Boulevard Allentown, PA 18195 1-800-345-3148	McWhorter Inc. 400 East Cottage Avenue Carpentersville, IL 60110 1-800-228-5635
Byk-Chemie USA 524 S. Cherry Street Wallingford, CT 06942 (203) 265-2086	Patco Specialty Chemicals 3947 Broadway Kansas City, MS 64111 1-800-821-2250
Cabot Corporation Cab-osil Division P O Box 188 Tuscola, IL 61593-0188 1-800-222-6745	Rohm & Haas Independence Mall West Philadelphia, PA 19105 1-800-523-4480
Daubert Chemical Co. S. Central Avenue Chicago, IL 60638 (708) 496-7350	S.C. Johnson Polymers 1525 Howe Street Racine, WI 53403-5011 1-800-231-7868
Drew Industrial Division One Drew Plaza Boonton, NJ 07005 (201) 263-7600	SCM Chemicals 7 St. Paul Street Suite 1010 Baltimore, MD 21202 1-800-638-3234
Dow Chemical Midland, MI 48674 1-800-447-4369	Union Carbide / Solvents & Ctgs. Mtls. Dv 39 Old Ridgebury Rd Danbury, CT 06817-0001 1-800-568-4000
DuPont 1007 Market Street Wilmington, DE 19898 1-800-441-9442	Witco Corporation Oleochemical/Surfactant Div. 520 Madison Avenue New York, NY 10022 (212) 605-3680
Eastman Chemical Co. Kingsport, TN 37662 1-800-EASTMAN	

Appendix II

Equipment	Manufacturer/ Supplier

Airless Spray:
Por*table* Hydra-Spray Supply Pump. Model 224-618 Series — Grayco
A containing a 45:1 Ratio King Pump. An ARO High
Pressure Material Regulator(downstream type) Model
651780-A2B is used to regulate fluid pressure.

Model 550 Automatic Airless Spray Gun — Binks
Twist-Tip II Reversible Nozzle Cleaner
1/4" No-wire Airless Fluid Hose

Spraymation Model 310410 Automatic Test Panel Spray — Spraymation
Machine

Test Panels:

ACT Cold Roll Steel 4X12X032. Unpolished. APR10161. — ACT
ACT Hot Roll Steel 3U 4X12X071. Blast Clean.
Unpolished. APR18567

Drawdown
Bird bar, 3" film width, to deliver 3.2 mils. dry.

Glossmeter:

Hunter Lab PRO-3 Glossmeter (meets ASTM D-523) — Hunter Lab

Viscosometers:

Stormer Viscometer, Krebs-Type — Byk-Gardner

Brookfield Viscometer Model LVTDV-II — Brookfield
I.C.I. Cone & Plate Viscometer — ICI
Rheometrics Recap III-Dynamic Mechanical Analyzer — Rheometrics

Corrosion Testing:

Salt-Fog Chamber Model GS-SCH-22 — Harshaw/Filtrol

continued…

Appendix II (Cont'd)

High Shear Foam Tester:

Commercial Waring Blendor, with 8 oz. stainless steel cup   Waring

Particle Size Measurement:

Lemont Scientific Image Analysis, Program AC-08251.   Lemont Scientific
Line Scan Image Analysis for Particles
Fineness of Grind Gage   Byk-Gardner
Stereomicroscope   Cambridge Instruments

Average Surface Roughness:

Tencor P-2 Long Scan Profiler   Tencor Instruments

Dry Time

Drying Time Recorder   Paul N. Gardner Co.

References

(1)   Levinson, S.B., *Application of Paints & Coatings*,Federation Series on
        Coatings Technology; Federation of Societies for Coatings Technology:
        Philadelphia, 1988.
(2)   Chigier, N.A., "The Physics of Atomization," *ICLASS-91*, Plenary Lecture,
        Gaithersburg, MD , July 1991, p 1-15.
(3)   Kwok, K.C. and Liu, B.Y.H., "How Atomization Affects Transfer Efficiency,"
        Industrial Finishing 1992, *68 (5)*, 28-32.
(4)   Lefebvre, A.H. and Senser, D.W., "Research Unravels Spray Mysteries," *Industrial
        Finishing* 1990, *66 (6)*, 16-20.
(5)   Bryant, D.A. and Nae, H.N., "The Effect of Associative Thickeners on the
        Properties of Water Based Coatings," presented at the 20th Water-Borne, Higher-
        Solids and Powder Coatings Symposium, New Orleans, February 1993.
(6)   Maver, T., "Rheology Modifiers: Modifying Their Performance in High Gloss
        Paints," *Journal of Coatings Technology* 1992 *64 (812)* , 45-57.
(7)   Schwartz, J., "The Importance of Low Dynamic Surface Tension in Water-Borne
        Coatings," *Journal of Coatings Technology* 1992 *64 (812)*, 65-74.

(8)   Ross, S. and Morrison, I.D., *Colloidal Systems and Interfaces,* John Wiley & Sons: New York, 1988, p 165-170.

(9)   Adamson, A.W., *Physical Chemistry of Surfaces,* 5th Edition, John Wiley & Sons: New York, 1990, p 550.

(10)   Ross, S. and Young, G.J., "Action of Antifoaming Agents at Optimum Concentrations," *Industrial & Engineering Chemistry* **1951,** *43 (1),* 2520-2525.

(11)   Ross, S. and Butler, J.N., "The Inhibition of Foaming. VII. Effects of Antifoaming Agents on Surface-Plastic Solutions," *Journal of Physical Chemistry* **1956,** *60 (8),* 1255-1258.

(12)   Rosen, M.J., *Surfactants and Interfacial Phenomena,* John Wiley & Sons: New York, 1978, p 191.

(13)   Ross, S. and Morrison, I.D., *Colloidal Systems and Interfaces,* John Wiley & Sons: New York, 1988, p 110-112.

(14)   Ibid, p 171-172.

(15)   Bendure, R.L., "Dynamic Surface Tension Determination with the Maximum Bubble Pressure Method," *Journal of Colloid and Interface Science* **1971,** *35 (2),* 238.

(16)   Adamson, A.W., *Physical Chemistry of Surfaces,* 4th Edition, John Wiley & Sons: New York, 1982, p 23-24.

(17)   Antonucci, E., "Foam Control in Water-Borne Coatings," *Journal of Water-Borne Coatings* **(1980),** 3 (2), 4.

Metallic Pigments for Waterbased Coatings

W. Reißer, A. Fetz, and E. Roth

SPARTE ALUMINIUMPIGMENTE, WERK GÜNTERSTHAL, ECKART-WERKE, D-91235 VELDEN, GERMANY

1 INTRODUCTION

The problem of using aluminium pigments for waterborne coatings is the exothermic reaction caused when bringing water and aluminium together.
This can be shown by the following equation 1:

$$2\ Al + 6\ H_2O \longrightarrow 2\ Al\,(OH)_3 + 3\ H_2 \tag{1}$$

When converting regular aluminium into the flake form of pigment the surface area is greatly increased, causing a major increase in surface reactivity with water.
Up until the 1970's it was almost impossible to use conventional aluminium pigments in ready mixed stable waterborne coatings, and it was therefore necessary to use a two pack system.
In the meantime pigment and paint manufacturers have succeeded in developing a wide range of storage stable one pack systems for a wide variety of coating applications.
These coatings are manufactured from specially prepared aluminium pigments that are even used in the demanding automotive industry where only the highest quality is acceptable.
Today production line coating capability with waterborne metallic paints is performed as a routine matter.[32]

2 STABILIZATION OF ALUMINIUM PIGMENTS

Aluminium pigments must not only be stable against water but also a wide variety of coating components. Gassing stability depends on both passivation technology and coating formulation and therefore both parameters must be considered.
The aim of the metallic pigment manufacturer is to develop aluminium pigments that exhibit sufficient stability in a wide variety of common binder systems. A simple hydrophobization of the pigment surface, for example by absorption of saturated fatty acids, is not suitable for this purpose.
Many methods have been considered for the effective stabilization of aluminium pigments and they can be divided into two categories:

(a)   The adsorption of corrosion inhibitors on the pigment surface.
(b)   The encapsulation of the pigment with either inorganic or organic (polymer) coatings.

2.1 Stabilizing of aluminium pigments with corrosion inhibitors

The stabilizing mechanism of aluminium pigments by corrosion inhibitors is by their chemical adsorption at the active sites on the pigment surface, which thereby inhibits the reaction of aluminium and water.

There are a number of methods described in various patent literature, including: phosphoric esters,[1-6] phosphates and phosphonites,[7-9] vanadates,[10-12] chromates,[13-15] molybdates[16] and dimeric acids[17] etc.

Some of the described methods and technologies can lead to the formation of coherent coatings on the pigment surface and therefore could be considered as polymer encapsulation.

2.2 Stabilizing of aluminium pigments by organic or inorganic protective coatings

The possibility to coat aluminium pigments with inert silica has been well known for some time.[18,19] This technology provides the resultant pigment with very good chemical and thermal resistance properties. The main applications of these pigments are in aqueous coatings, powder coatings, electrostatically applied coatings and thermoplastic masterbatches.

Other inorganic coatings such as iron oxide[20] or titanium dioxide[21] provide new and unique effect pigments. These products need additional stabilization to render them suitable for aqueous coatings.

Patent literature describing polymer encapsulation are mainly in the phosphoric acid modified polymer group using acids as the coupling agent between the inorganic surface and the polymer.[22-27]

Other polymer groups have been used to obtain the gassing stability required when producing aqueous coatings.[28-31]

3 RESIN AND ALUMINIUM PIGMENT SELECTION

Even with state of the art development it still takes serious technical knowledge and ability to formulate stable water based coatings.

An important factor to consider is the pH of the coating composition as aluminium is amphoteric, and is therefore most stable at pH 7. Reactivity will increase rapidly with both acidity and alkalinity and therefore these parameters have to be examined as most paint resins in use are acidic with neutralisation achieved by addition of strong alkalis. Dispersion (emulsions) binder systems are normally less critical as the pH range normally falls between 5 and 8. Solution resin types are normally more critical with the pH often above 8 making the system very aggressive to aluminium pigments.

In general the resins should be neutralized with base before addition of the aluminium premix. Milder amines such as 2-dimethylaminoethanol or 2-amino-2-methyl-1-propanol work best. Ammonia is very aggressive and can cause gassing and therefore should be avoided.

The most important ciriteria for aluminium pigment selection is the extent of passivation that is necessary to achieve a stable paint system. Usually the stability of an inhibited pigment is lower than that of an encapsulated pigment. However the inhibited pigments are normally cheaper and have superior optical properties.

4 APPLICATION EXAMPLES

Typical applications for waterborne aluminium pigments are anticorrosive paints, primers, roof coatings and printing inks. Even coatings for the automotive industry use these products including OEM, repair and plastic accessories. The following examples demonstrate that commercially available aluminium pigments are suitable for formulating superior waterborne coatings.

Four different pigments were selected all having different stabilizing technologies (Table 1). The stabilization of the pigments AC and CR/PM was performed according to[27] and[14] respectively.

To achieve objective results all pigments were prepared from the same medium fine aluminium pigment (laser granulometer: D10 10 microns, D50 17 microns, D90 26 microns). This is because particle size and particle size distribution as well as chemical treatment significantly effect the end product.

Table 1   *Stabilized Aluminium Pigments*

Pigment Type	*Passivation*	*Solvent*
SI	SiO_2 encapsulation	-
P01/PM	Phosphorous organic treatment	35 % Methoxypropanol
AC	Acrylic polymer encapsulation	35 % White Spirit
CR/PM	Chrome treatment	35 % Methoxypropanol

4.1 Automotive two coat metallics

The pigments P01/PM, AC and CR/PM were evaluated in two different basecoats, clearcoat formulations for the following criteria:

- Gassing stability
- Flop / two tone *
- Distinctivness of image (DOI) *
- Substrate adhesion *
- Intercoat adhesion *
- Circulation stability *
    *   Before and after humidity test (according to DIN 50017)

4.1.1 Gassing Stability.   There were several different methods published,[31, 32] but the only one suitable for the purpose which was relatively reproducable is the measurement of gassing during storage of the basecoat at 40 °C (40 °C-test). This test is completed when 25 ml of hydrogen are generated within 30 days exposure time.

Although the 40 ^{o}C test is very time consuming it is becoming the most favoured method.[32]

Test results of two different basecoats are shown in Figure 1.

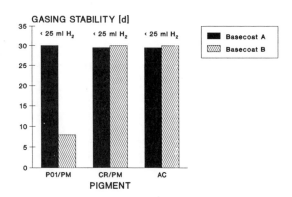

Figure 1   *40 ^{o}C Gassing Test of the Stabilized Aluminium Pigments in the Basecoats A and B*

In basecoat A (aliphatic polyurethane dispersion) all three pigments performed well, however in basecoat B (polyester melamine system) both the chrome treated and the polymer encapsulated pigment showed good gassing stability. The phosphorous organic treated pigment P01/PM exhibited a less effective passivation, and therefore appears to be less stable in this special formulation.

These results are also evidence of the already mentioned dependence of gassing stability on both binder formulation and pigment passivation.

4.1.2 Application Properties.   Flop, DOI, substrate and intercoat adhesion stability were all tested in a basecoat/clearcoat system on phosphated steel panels as follows:

- Cataphoretic primer
- Filler (25 - 30 microns)
- Metallic basecoat (dry film thickness 12 - 15 microns)
  Intermediate drying 5 min/60 ^{o}C
- Clearcoat (acryl melamine, dry film thickness 30 - 35 microns)
  Flash off time approx. 10 min
  Baking 20 min/140 ^{o}C

Optical properties.

Flop values (determined with a Zeiss goniophotometer GP 3) and DOI (determined with the Dorigon D 47 R 6F, Hunter) were checked before and after 240 hours humidity test (DIN 50017).

Flop evaluations are shown in Figure 2, DOI readings in Figure 3.

The flop characteristics of the chrome treated pigment are slightly poorer than that of the polymer encapsulated and the inhibited types.

In all cases condensation water exposure lead to a slight flop reduction as is normal with conventional coatings. Before humidity testing the DOI values were similar for all the products, but there was a remarkable reduction after 240 hours exposure, the chrome treated pigment being the best, followed by the polymer encapsulated type.

Substrate Adhesion.

The cross cut test (DIN 53151) shows all the pigments tested exhibit the same behaviour before and after 240 hours condensation water test (GT 0).

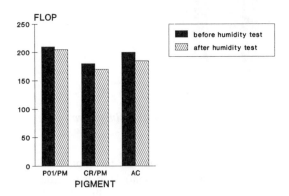

Figure 2   *Flop-Values before and after 240 h Humidity Test of Metallic 2-Coat Paint Applications (Basecoat A)*

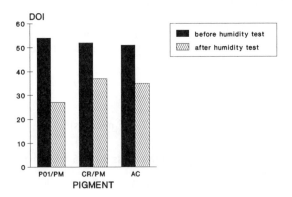

Figure 3   *DOI-Values before and after 240 h Humidity Test of Metallic 2-Coat Paint Applications (Basecoat A)*

Intercoat Adhesion.

The intercoat adhesion was tested with a gravelometer according to the Ford specification, the results are given in Figure 4.

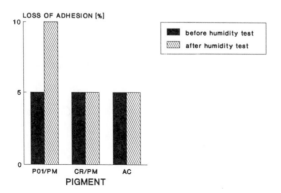

Figure 4   *Gravelometer Test according to the Ford-Specification before and after 240 h Humidity Test of Metallic 2-Coat Paint Applications (Basecoat A)*

Both the chrome treated and the polymer encapsulated pigment show excellent performance after 240 hours exposure. The inhibited pigment shows a clearly measurable deterioration.

Circulation stability.

The circulation stability was tested with the following methods (Table 2):

Table 2   *Test Methods for Circulation Stability*

Method	Testing Time
Waring blender	3 min
Ultrasonic test	20 min
Gearwheel pump	60 min
Circular pipeline	2 weeks

The mechanical stress of these different testing methods had practically no detrimental influence on optical characteristics, gassing stability, adhesion and condensation water resistance of the evaluated pigments or subsequent coatings.

4.2 Coatings for plastic accessories (two coat)

The following formulation is a two coat metallic for plastic accessories which is particularly suitable for hub caps.

The passivated pigments SI, P01/PM, AC and CR/PM were used in the following coating system on polyamide (pre-cleaned with white spirit):

- Metallic basecoat (acrylic polyurethane dispersion), dry film thickness 10 - 20 microns and 5 min intermediate drying at 60 °C.
- Clearcoat (medium solids acrylic polyurethane), dry film thickness 30 microns, baked for 30 min at 60 - 80 °C.

Flop and substrate adhesion (cross cut test) before and after 240 hours humidity test are shown in Table 3.

Table 3   *Flop Values and Cross Cut Test Results before and after 240 h Humidity Stress*

Pigment	Flop		Cross Cut Test	
	before 240 h	after 240 h	before 240 h	after 240 h
SI	217	215	GT 0	GT 0
P01/PM	233	234	GT 0	GT 1
AC	158	159	GT 0	GT 0
CR/PM	280	278	GT 0	GT 0

The CR/PM pigment has the best flop characteristics, followed by P01/PM, SI and AC. Condensation water exposure has no influence on flop or substrate adhesion with one exception: The cross cut test result of the P01/PM product is slightly poorer (result GT 1), but is still acceptable.

In addition, blistering according to DIN 53209 was measured with excellent results for all formulations (m 0/g 0).

4.3 Coating for plastics (one coat)

The following example of a phono equipment coating describes the application properties of a one coat metallic based on a styrene acrylic resin system.

Polystyrol (pre-cleaned with white spirit) was coated with a styrene acrylic dispersion containing the pigments SI, P01/PM, AC and CR/PM. The coating had a dry film thickness of 15 - 20 microns after baking for 40 min at 60 °C.

For this application product resistance against household cleaners and hand cream as well as substrate adhesion are very important. Table 4 shows chemical resistance against tap water, ethanol (60 % in water), white spirit and Nivea hand cream.

Table 4   *Resistance against Tap Water, Ethanol, White Spirit and Nivea Cream.*

Pigment	Tap Water a)	Ethanol b)	White Spirit c)	Nivea d)
SI	1	2	1	1
P01/PM	1	3	1	3
AC	1	2	1	2
CR/PM	1	2	1	2

Tests: *b) 15 min Ethanol exposure under watch glass, a) + c) 50 rub cycles with tap water or white spirit soaked cotton pad, d) Coating with Nivea cream 3 days at 50 °C;*

1 no change, 2 very slight tarnishing, 3 slight tarnishing, 4 strong tarnish, 5 very strong tarnish.

Table 4 shows that the best results were achieved with the pigment SI followed by AC and CR/PM, with the P01/PM giving unsatisfactory results with this specific paint system.

Examination of substrate adhesion by the cross cut test and tape test is shown in Table 5.

Table 5 *Tape Test and Cross Cut Test Results*

Pigment	Tape Test	Cross Cut Test
SI	No change	GT 1
P01/PM	No change	GT 1
AC	No change	GT 1
CR/PM	No change	GT 1

4.5 Can/Coil Coating

The pigments SI, P01/PM, AC and PR/PM were all tested in coatings based on a polyester melamine system, applied to cold rolled steel (pre-cleaned with ethylacetate). This was baked for 90 min at 260 °C, giving a dry film thickness of approximately 12 microns.

Flop, substrate adhesion (cross cut test) and humidity resistance were all tested. Figure 5 shows that before the 240 hours water exposure test the coating containing the pigment P01/PM was the best, followed by the coatings with SI and AC. However the coating with the P01/PM pigment has a very strong decrease in flop value after the water condensation exposure. Therefore this pigment would not normally be recommended for this binder system. As a better choice would be both the SI and AC types.

Excellent results were obtained with the test methods used for blistering and adhesion (cross cut test) before and after the 240 hours humidity test (m 0/g 0 and GT 0 respectively).

5 CONCLUSION

The results presented clearly demonstrate that the different chemistries and technologies available today make it perfectly possible to manufacture waterborne coatings containing aluminium pigments for a wide field of applications.

These said pigments are therefore an important contribution to the ecological aim of reducing the levels of VOC's in both the production and application of modern coatings.

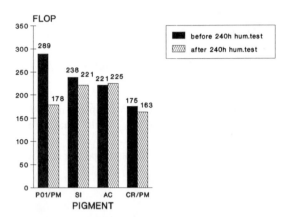

Figure 5   *Flop Properties before and after 240 h Humidity Test*

References

1   DEP 3020073
2   EP 853051480
3   EP 0170474
4   EP 841073661
5   USP 4565716
6   USP 4808231
7   USP 4565716
8   USP 4808231
9   EP 0240367
10   EP 0305560
11   EP 0104075
12   USP 5215579
13   USP 2904523
14   USP 4693754
15   EP 0259592
16   EP 0583919
17   DEP 3002175
18   USP 2885366
19   USP 3954496
20   DEP 3003352
21   DEP 3813335
22   EP 0170474
23   CAP 1273733
24   ATP 372696
25   DEP 3807588
26   EP 0319971
27   US 5332767
28   DEP 3630356

29 DEP 3147177
30 EP 0477433
31 B. Müller and G. Niederberger, Farbe und Lack, 1993, **99**, 995.
32 R. Besold, W. Reißer and E. Roth, Farbe und Lack, 1991, **97**, 311.

Polymeric Surfactants and Their Application in Resin Emulsification

A. Bouvy and A. Opstaele

ICI SURFACTANTS, PAINTS AND COATINGS, EVERSLAAN 45, B-3078 EVERBERG, BELGIUM

1 INTRODUCTION

Surfactants are typical performance chemicals that are essential in developing many successful formulations. They find application in almost every industrial area and impact every aspect of our daily lives - from the food we eat to pharmaceutical formulations that make our lives longer and more comfortable [1].

Surfactants, or surface active agents, have the ability to lower the surface tension of the air-water or water-oil interface and also, in general, they strongly adsorb at the solid-liquid interphase. A direct consequence of adsorption at interfaces is that the corresponding surface free energy is significantly reduced and this results in surface active effects such as foaming, wetting, emulsification and dispersion [2].

Nowadays, there is an increasing awareness that surfactants are structural and functional entities of considerable complexity, where effects can be optimised by molecular design and application technology. It is from this kind of scientific approach that polymeric surfactants have emerged and they are an exciting new addition to the large variety of surfactants already available [1,3-5].

2 STABILISATION MECHANISMS

The use of ionic (usually anionic) surfactants to stabilise colloidal particles by electrostatic interactions is well-known nowadays and the DLVO theory provided a functional guidance to formulation work with them [6]. Coupled to the fact that ionic surfactants were available for a number of years before the appearance of ethoxylated products and that they are often more cost-effective in less demanding conditions for emulsions or dispersions when compared with nonionic surfactants, the use of ionics is still widespread.

At a time when formulation chemists are demanding much broader applicability of the surfactants, when formulations are becoming even more complex, it is the steric stabilisation phenomena, induced by nonionic surfactants, that are becoming of more importance.

The electrostatic stabilisation mechanism is highly effective but it is restricted to media of high dielectric constant. However, steric stabilisation offers several distinct advantages.[7]. These are briefly undermentioned :

- Equally effective in aqueous and non-aqueous media
- Production of dispersions in extreme conditions of dispersed phase volume, polarity of dispersion medium, temperature, ionic strength
- Higher flexibility in designing tailor-made molecules for maximum performance and cost effectiveness

3 POLYMERIC SURFACTANTS AND STERIC STABILISATION MECHANISM

It has been shown experimentally that the best steric stabilisers are amphipathic in character. This is particularly true with high molecular weight species that take the form of block, random or graft copolymers. These molecules are built up from at least two chemically bound groups : one of which is soluble in the dispersion medium (stabilising moiety) whilst the other is nominally insoluble (anchoring group) [7,8,9].

3.1 Stabilising Moiety

The main requirement for the stabilising moieties is that they must have appreciable solubility in the continuous phase and therefore strongly solvated by the medium. The soluble chains - which project away from the particle surface into the dispersion medium - are responsible for the observed stability by preventing particles from approaching too closely to each other. This function is better fulfilled when chain-solvent interactions are stronger than chain-chain interactions between two similarly protected particles.

As particles approach each other in a Brownian collision, any interpenetration of the stabilising layers will cause solvent molecules to diffuse into the overlapping region, hence pulling the particles apart. In contrast, if the inter-chains attraction prevails, flocculation will inevitably occur.

An additional requirement for the soluble chains is that they must achieve a complete surface coverage for the dispersed particles and create a stabilising layer of sufficient thickness to prevent particles approaching too closely.

3.2 Anchoring Moiety

Because of the affinity for the continuous phase, the stabilising chains cannot adsorb strongly enough to the surface of the dispersed phase particles, thus ensuring that no desorption or displacement occurs on collision. Therefore anchoring groups are required to be present in the stabilising molecule to fulfil this function.

When the dispersed phase is a liquid, the anchoring group, as well as being insoluble in the continuous phase, should also be fully soluble or at least compatible with the dispersed phase.

Apart from the requirements for solubility, it is obvious that the stabilising molecule must contain a sufficient number of anchoring points and that the mechanism of anchorage must take into account the stress that the stabilised system is expected to withstand. This is particularly achieved with polymeric species because the greater number of anchoring groups per molecule makes the total adsorption energy high, even if individual contacts are weak.

The foregoing is not meant to imply that homopolymers cannot impart steric stabilisation [7]. Homopolymers are relatively ineffectual, however, because of the conflicting requirements that the dispersion medium be a poor solvent to ensure strong adsorption of the stabilizer onto the particle, but a good solvent to impart effective steric stabilisation. As a result the instability of dispersions stabilized by homopolymers often arises from the lateral movement and/or desorption of the stabilizing chains.

4 POLYMERIC SURFACTANT STRUCTURES

In principle, there is an almost infinite number of polymeric structures suitable as sterically stabilising surfactants. However, as illustrated in Figure 1, three basic structures seem to be most relevant in a large variety of industrial applications and these are now described in some detail.

4.1 "Random" Polymeric Structures

These are random, three-dimensional networks synthesised from polyalkylene glycols, polyols, aliphatic carboxylic acids, aliphatic and/or aromatic polycarboxylic acids or anhydrides. The resulting products are statistical mixtures, with a wide molecular weight distribution, that can be described to the first approximation as containing loops of hydrophilic moieties (e.g. polyoxyethylene) bonded through ester linkage to the lipophilic moieties.

By changing the proportion and type of components it is possible to produce a broad range of surfactants with different characteristics, such as water or hydrocarbon solubility, cloud point (for the water-soluble polymers), emulsion stabilisation at high temperature, ionic strength and shear dispersion properties in aqueous and non-aqueous media.

The key features of the random type of polymeric surfactants are that they exhibit both emulsification and dispersion properties, and that they impart good resistance to emulsion coalescence even at low interfacial coverage and at high shear.

4.2 "Ordinate" polymeric structures

This A-B-A type of polymer - firstly commercialised by Wyandotte Chemicals Corporation [10] - is probably the most well-known and the most extensively studied amongst the family of polymerics [11]. It consists of a polymerised propylene oxide central segment onto which ethylene oxide is block copolymerised providing the potential for a very large range of related surfactants. The molecular weight and block composition of the copolymers determine hydrophilic/hydrophobic balance and control the surfactant properties.

In order to increase the difference in the polarity and solubility characteristics of the ethylene oxide/propylene oxide moieties and consequently to extend their application area, a new class of block polymers has been more recently developed by ICI. These compounds have been obtained by esterification of poly(12-hydrostearic acid), (PHSA), with polyalkylene glycols to give an A-B-A type block copolymer [12].

Their structure is conceptually similar to the classical ethylene oxide/propylene oxide condensates. Not only the differences in above mentioned characteristics of segments A and B are much more pronounced but also the hydrophobic PHSA moiety is more soluble than propylene oxide in a variety of oil phases.

In O/W emulsions the hydrophobic chains function as anchoring groups and the interactive stabilisation with the dispersion medium is ensured by the hydrophilic centre of high molecular weight polyalkylene glycol.

The polarity of ordinate polymeric surfactants can be tailored by varying the PHSA content, or the molecular weight and composition of the alkoxylated core in order to achieve the desired degree of interaction with either oily or aqueous phases of different composition.

Molecules from this family have a remarkable versatility; they are suitable to formulate both O/W and W/O emulsions, as well as to disperse pigments in aqueous and non-aqueous media.

4.3 "Comb" Type Structure

If any polymeric surfactant has an "ideal" structure then this class of polymeric surfactants comes very close.

The concept of a relatively ordinate, repetitive structure with a multiplicity of anchoring and stabilising units has been realised industrially with the production of "comb" type molecules, where an acrylic/methacrylic backbone provides the support to hydrophilic pendant groups based on polyalkylene glycols.

Again a large number of molecular parameters can be varied, from the molecular weight and composition of the backbone to the number of chains per unit length of the backbone, to the molecular weight and composition of the chains.

A "comb"-type structure lends itself best to a fine tuning of the molecule to optimise both the chemical and steric interaction and achieve maximum interfacial effects, particularly in dispersing solids in water.

Figure 1 : *Schematic representation of some typical structures of Polymeric Surfactants*

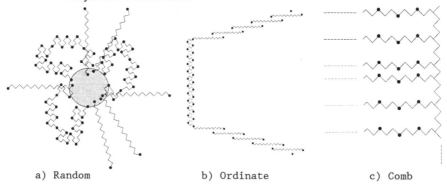

a) Random   b) Ordinate   c) Comb

5 ADVANTAGE OF POLYMERIC SURFACTANTS

The effectiveness of a surface active agent in stabilising dispersions depends on the extent and intensity of the interaction between its amphipathic groups and the dispersed phase and the continuous medium. In the classical chemistry of surface active agents, ionic or nonionic polar groups and hydrophobes of C_{12}-C_{18} chain length are combined to produce molecules having one single interaction point with each of the immiscible phases. These molecules have an adequate level of surface activity for aqueous systems and for application in non-demanding conditions.

The situation is quite different when more severe requirements are involved, for example, because of non-aqueous environments or because of temperature, shear, ionic strength, dispersed phase volume ratio, polarity and molecular weight of the oil component. In these instances it is essential to enhance the strength of the interaction. One most practical and effective possibility is to increase the number of interacting sites per molecule i.e. to use polymeric surfactants.

It is generally acknowledged that polymeric species give effective stabilisation provided they satisfy the conditions previously mentioned, namely

(a) Complete coverage of the surface of the dispersed particles to prevent even transient contacts between unprotected areas

(b) No desorption of stabilising chains during particle collision

(c) Maximum number of configurations possible for the stabilising species in a non-collision situation

(d) Good solvation of the stabilising chains by the continuous medium

(e) Thick adsorbed layer to start repulsion at large distance between approaching droplets.

The first two conditions imply that the adsorption energies for the stabilising molecule must be large. This is better achieved with polymeric molecules as explained previously. It is also more difficult, statistically, to desorb a molecule attached through multiple anchoring points than one adhering through a single linkage.

A high number of possible configurations is required to maximise the entropic contribution to the total repulsion energy that prevents particles from coalescing. This increases rapidly with molecular size. Small molecules may be regarded as essentially non-flexible, and only their orientations relative to the surface count as different conformations. With polymeric materials additional conformation possibilities arise from the relative size of the loops and, because of their comparatively large dimensions, from the actual steric arrangements of each loop.

Finally, provided that particles are well covered by thick adsorbed layers i.e. large stabilising chains, fully extended in the medium, dispersions approaching thermodynamic stability-rather than colloidal stability - can be produced for dilute systems [4,7]. This will never be obtained by using classical surfactants - anionics or even nonionics.

6 APPLICATION OF POLYMERIC SURFACTANTS IN THE PRODUCTION OF AQUEOUS PAINTS

6.1 Background

Steric stabilisation with ICI polymeric surfactants has already been applied in the preparation of oil or solvent in water emulsions [3,13] multiple emulsions [5,14], suspoemulsions or suspension concentrates in agrochemical formulations [15,16]. Their use in inverse emulsion polymerisation of acrylamide is also documented [17].

This concept is now extended to prepare resin emulsions such as alkyds or epoxies for the production of aqueous paint formulations.

The most extensively used water-borne paints today are based on polymer emulsions, typically prepared in-situ polymerisation in the aqueous phase. The advantage of this technology (in formulating aqueous paints) is that the binder is already dispersed in the same solvent as the final paint. In some applications, however, polymer emulsion-based paints are not entirely satisfactory, for example, when high gloss or good penetration into porous substrats are needed. In these cases, organic solvent-borne alkyd paints are still used, despite growing environmental concerns.

Provided that typical drawbacks of emulsions such as colloidal stability can be overcome, oligomeric resin emulsions are a potentially attractive - but challenging - alternative to both polymer emulsion and solvent borne paints as confirmed by several recent publications [18-23].

6.2 Preparation and Stability of Emulsions

The purpose of the emulsification process is to create numerous small droplets from a large homogeneous oil phase (the resin) through the application of surfactants and input energy. The essential characteristics of the resulting emulsions are :

- The emulsion type : O/W or W/O. This is primarily determined by the type of surfactant as explained by Bancroft rule [24].
- The droplet size distribution, which strongly influenced the emulsion stability. However, if it is easy to make droplets, it may be more difficult to make them small enough and distributed uniformally [25].

Because of the large interfacial area created during the emulsification process, emulsions are thermodynamically unstable and subject to change over time. Obviously, this poses major concerns to a formulator because of the required shelf life of the paint. Surfactants will not only facilitate size-reduction by lowering interfacial tension between resin and water phases but they will also impart long term and mechanical stabilities to the emulsion [26].

Several methods can be used to prepare emulsions and to assess their stability. Plenty of very well detailed reviews, covering all aspects of emulsion technology, are available from literature [2,27-29].

The technique used to prepare resin emulsions is referred as Emulsion Inversion Point (EIP) and consists in gradually increasing the water concentration added into the oil phase (resin + surfactant) at constant temperature. At first stage, a W/O emulsion is obtained which breaks up to form (W/O)/W type double emulsion before it completely inverts into the intended O/W emulsion. The EIP is the point at which the emulsion changes from W/O to an O/W system and therefore represents the number of cubic centimeters of added water per cubic centimeter of the oil present when the phase inversion occurs [28].

6.3 Surfactant selection

Providing that appropriate surfactant and emulsification conditions have been correctly selected in function of resin type and characteristics, most of the traditional air-drying alkyds and liquid epoxies can be succesfully emulsified with the E.I.P. method.

Table 1 presents both the type of resins investigated and the recommended surfactants. To obtain fine emulsions, it is advantageous to add emulsifier to the resin phase because of the lower interfacial tension obtained than when added into the water phase prior to surfactant distribution between oil and water phases attaining equilibrium [28].

Emulsions of long oil alkyds and epoxies (EEW inferior to 515) are prepared without any solvent and contain up to 60 % resin phase. Higher viscous resins - such as medium and short oil alkyds as well as solid epoxies - require the presence of solvent in variable quantities as illustrated in Table 1.

Table 1 *Polymeric Surfactant Selector Chart*

Resin type	Solvent	Surfactant	Structure
Alkyd			
Long oil	-	'Atsurf' 3969	Block
Medium oil	3 % Methoxy-Propanol	'Atsurf' 3969	Block
Short oil	50 % Xylene	'Atsurf' 3109/'Atsurf' 3300B	Random
Epoxy			
165 < EEW < 180	-	'Atlas' SCS 2721	Blend (*)
185 < EEW < 190	-	'Atsurf' 3350	Random
EEW $\simeq$ 250	-	'Atlas' SCS 2724	Blend (*)
450 < EEW < 515	-	'Atsurf' 707	Block
1550 < EEW < 2000	25 % Butoxyethanol	'Atsurf' 108	Block

(*) Experimental product

6.4 Alkyd Emulsion Properties and Paint Performances

The influence of emulsification conditions - including surfactant concentration - on both emulsion and paint properties has been investigated using starting formulations.

Table 2 presents a typical white gloss paints based on a medium soya oil length emulsified with 'Atsurf' 3969. More practical details on the preparation of the resin emulsion and paint formulation can be found in references 30 (alkyds) or 31 (epoxies).
Basic properties of both emulsion and paint are summarised in Table 3.

Polymeric surfactants are also very effective in dispersing pigments. The multi-anchoring group effect will cause the surfactant to be strongly adsorbed onto the pigment surface, reducing desorption during the letdown and therefore allowing better compatibility with the resin emulsion.

In the emulsion technology, increased surfactant concentration may facilitate at least to some extent the production of a very fine and stable emulsion with narrow droplet size distribution. It is actually important that, in the emulsification process, newly created interfaces are rapidly and completely covered by surfactant molecules, once being formed. This is directly dependent on the availability of surfactant. A too high surfactant concentration will, however, cause a reduction in the water resistance of the paint film or increase its drying time and stickness. Costs involved are also difficult to afford.

Table 2 *Air drying high gloss decorative building paint*

Components	% W/W	Properties/Suppliers
Nebores SP 29 [1]	32.4	VOC : 99 g/l
1-methoxy-2-Propanol [2]	1.6	PVC : 15.2
'Atsurf' 3969 [6]	2.3	Solid content : 58 %
Nuodex Co (8 %) [3]	0.12	
Nuodex Ca (10 %) [3]	0.8	(1) Necarbo B.V.
Nuodex Zr (12 %) [3]	1.27	(2) Merck
Additol XL 297 [4]	1.0	(3) Servo Delden BV
Tioxide TR 92 [5]	23.0	(4) Hoechst
Atlas SCS 2102 (20 %) [6]	0.92	(5) ICI Tioxide
Hypermer CG6 (32 %) [6]	3.0	(6) ICI Surfactants
Antifoam	q.s.	
Water	up to 100	

Table 3 *Emulsion and paint properties*

Properties	Initial	1 Year
Emulsion		
* **Stability**		
(RT)	-	No change
* **Particle size**		
x = (μm)	0.7	No change
90 % <	1.6	No change
* **Viscosity**		
(mPa.s)	960	1000
* **pH**	3	3
Paint		
* **Stability**		
(RT)	No change	No change
* **Gloss**		(1) (2)
20°	82	71 71
60°	93	91 88
* **pH**	6	6
* **Drying time**		
1 Hr	< 2	< 2
4 Hrs	4	4
7 Hrs	6	6
* **Viscosity**		
(mPa.s)	1500	1700
* **Water resistance**	No change	No change

with (1) = Daylight exposure
     (2) = In can storage

Table 4 presents the influence of 'Atsurf' 3969 on both emulsion and paint formulation properties. A minimum of 5 % surfactant on resin is necessary to ensure good emulsion stability and small particle size. At lower concentration, not only is the emulsion not stable but it is also incompletely inverted and large multiple emulsion droplets are observed.

Interesting to note that the drying time (measured by the Bandow Wolf Method DIN 53130) on a 100 μm wet film thickness is not significantly influenced by surfactant concentration. This is also reported in the literature. The investigated nonionic emulsifiers do not affect the distribution of the driers between the alkyd and aqueous phases of the emulsion and only gave a minor reduction in the drying performance [22].

Table 4 Influence of the 'Atsurf' 3969 concentration on emulsion and paint properties

'Atsurf' 3969 (% W/W on resin)	3		5		7		10	
EMULSION (60 % Resin content)								
* Viscosity (mPa.s)	85		160		960		1000	
* Particle size (μm): x = / 90% <	2.2 / 8.6		1.3 / 3.0		0.7 / 1.6		0.5 / 0.9	
* Stability 1 Month: RT / 40°C	Coagulation / Caking		No Change / S(R)		No change / Small S(R)		No change / Small S(R)	
PAINT	Initial	1 Month	Initial	1 Month	Initial	1 Month	Initial	1 Month
* Gloss 20° / 60°	82 / 93	45 / 83	82 / 93	72 / 91	82 / 93	72 / 92	82 / 93	75 / 92
* Drying time 1 Hr / 4 Hrs / 7 Hrs	<2 / 4 / 6		<2 / 4 / 6		<2 / 4 / 6		<2 / 4 / 6	
* Yellowing (*)	3		4		5		5	
* Covering (*)	3		4		5		5	
* Viscosity (mPa.s)	150		200		1500		2000	
* Stability 1 Month: RT / 40°C	Caking / Caking		No change / Small coagulation		No change / Small coagulation		No change / Small S(R)	
* PVC	15.2		15.2		15.2		15.2	

with (*) = 1 = Worse, 5 = Best
S(R) = Reversible Separation

Water resistance of the paint film is also unaffected.
Thanks to this high molecular weight and multianchoring group functionality - resulting in strong adsorption on interfaces - drawbacks of traditional low molecular weight/highly diffusing surfactant molecules are significantly reduced if not totally eliminated.

As in most emulsification experiments, any variation in experimental conditions will influence the properties of the prepared emulsions. To obtain small droplets at low emulsifier concentration, it is therefore essential to optimise parameters such as T° of emulsification and particularly when using nonionic surfactants [28].

Results indicate that - for a given alkyd - there is an optimum emulsification temperature i.e. giving the smallest particle size and narrow distribution.

Short oil alkyds are particularly difficult to emulsify - even with the original solvent. A combination of a polymeric surfactant ('Atsurf' 3109) with a primary emulsifier ('Atsurf' 3300 B) is required to obtain the optimum balance between interfacial tension reduction effect and stabilisation of the emulsion. Emulsion properties are directly influenced by the ratio between the two surfactants. Although it has to be optimised as a function of the resin type, in general, best emulsion properties - low viscosity and small particle size - have been found for a ratio ('Atsurf' 3109/'Atsurf' 3300 B) around 1/3. A guide recipe for developing an aqueous air drying grey machinery finish enamel can be found in reference 30.

6.5 Epoxy Resin Emulsification

The technology of resin emulsification has also been extended to low molecular weight epoxies. However, much less data have been produced to date compared to alkyds. Nevertheless, initial results look encouraging and justify further exploration [31]. Here again, correct selection of both surfactants and emulsification conditions are critical. As observed with alkyds, a typical surfactant concentration around 7 % on resin is necessary to ensure long term stability. Resin droplets formed using the recommended procedures are of 1 μm diameter or less and a resin content of 55 to 60 % can be achieved.

7 CONCLUSION

The move from solvent to water-based paint formulations provides a good but challenging opportunity to use polymeric surfactants for the emulsification of oligomeric resins. Thanks to their higher molecular weight, typical drawbacks of highly diffusing surfactant molecules are significantly reduced if not totally eliminated.
Providing that the appropriate surfactant is selected and emulsification conditions are well adapted to the resin characteristics, polymeric surfactants do allow the formulation of aqueous paints with properties that can sometimes be very similar to those of traditional solvent-based formulations.

REFERENCES

1.   M.R. Porter, Handbook of Surfactants, Blackie & Son Ltd,
      London (1991).
2.   M.J. Rosen, Surfactants and Interfacial Phenomena (2nd ed.),
      Wiley-Interscience, Chischester (1989).
3.   A.S. Baker, *Paint & Resin*, 37, March/April (1987).
4.   I. Piirma, Polymeric Surfactants, Surfactant Science Series
      (Vol. 42), Marcel Dekker Inc., New York (1992).
5.   G. Bognolo, *Specialty Chemicals*, June 1990, 232.
6.   T. Sato and R. Ruch, Stabilization of Colloidal Dispersion by
      Polymer Adsorption, Surfactant Science Series (Vol. 9), Marcel
      Dekker Inc, New York (1980).
7.   D.H. Napper, Polymer Stabilization of Colloidal Dispersions,
      Academic Press, New York (1983).
8.   K.E.J. Barrett, Dispersion Polymerization in Organic Media,
      Wiley London (1975).
9.   B.A., De L. Costello, P.F. Luckham, TH.F. Tadros,
      J. Colloid Interface Sci., (1992), **152**, 237.
10.   L.G. Lundsted, (assigned to Wyandotte Chemicals Corp.)
      US 267 46 19 filed April 6, 1954.
11.   P. Bahadur, G. Riess, *Tenside Surf. Det.* (1991), **28**, 173.
12.   European patent 0 000 424, filed June 23, 1978.
13.   ICI Surfactants, Technical Brochure 20-44E. (June 1993)
14.   Ph. Loll, M.C. Taelman, ICI Surfactants Brochure RP 112/94 E.
15.   G. Bognolo et al., ICI Surfactants Brochure RP 55/90E.
16.   Th. F. Tadros et al : 'Advances in Pesticide Formulation Technology'
      ACS Symp. Ser. (1984), **254**, 11.
17.   G. Bognolo, ICI Surfactants Brochure RP 30/89 E.
18.   A. Hofland, *Polym. Paint Colour J.* (1994), **184**, (4353), 350
19.   A. Hofland, F.J. Schaap, *Farg och Lack Scandinavia* (1990), **9**, 182.
20.   G. Ostberg, B. Bergenstahl, K. Sorenssen, *J. of Coatings Technology*
      (1992), **64**, (814), 33.
21.   G. Ostberg, B. Bergenstahl, M. Hulden, *J. of Coatings Technology*
      (1994), **66**, (832), 37
22.   G. Ostberg, B. Bergenstahl, M. Hulden *Colloids and Surfaces*
      (Accepted).
23.   T. Fjeldberg, *J. Oil & Colour Chemists' Assoc.*, (1987), **10**, 278.
24.   W.D. Bancroft, Applied Colloid Chemistry, Mc Craw-Hill New York
      (1928).
25.   P. Walstra, *Chemical Engineering Sci.* (1993), **48**, 333.
26.   Th. F. Tadros, *L'actualité Chimique*, Mai-Juin 1991, 167.
27.   P. Becker, Encyclopedia of Emulsion Technology, Marcel Dekker,
      New York (1983).
28.   L. Marszall, in Nonionic Surfactants, Surfactants Science Series
      (Vol 23), Marcel Dekker Inc. New York (1987)
29.   D. Myers, Surfactant Science and Technology, VCH Publishers Inc.
      New York (1988).
30.   ICI Surfactants : Technical Brochure : 64-21E.
31.   ICI Surfactants : Technical Brochure : 64-23E.

Microbiological Protection of Waterborne Paint Formulations

John W. Gillatt

THOR CHEMICALS (UK) LIMITED, EARL ROAD, CHEADLE HULME, CHESHIRE SK8 6QP, UK

PRESENT ADDRESS: THOR CHEMICALS JAPAN LIMITED, 19-17 KASUGA-CHO, IZUMIOTSU, OSAKA 595, JAPAN

1. INTRODUCTION

Water based surface coating formulations, most commonly referred to as paints, are susceptible to attack by microorganisms both in the wet state and, after application to a substrate, on the dry film.

During production and in the can growth of microorganisms can result in viscosity loss, gassing, malodour, pH drift, visible surface growth and other undesirable effects.

The dry film may become infected with fungi and algae resulting in disfigurement and breakdown of the coating followed by deterioration of the underlying substrate if that too is susceptible to such organisms.

The trend in the paint industry from solvent borne to water based formulations has led to greater demands for effective measures to prevent microbiological attack.

However, the use of chemical preservatives, or biocides, alone is not the complete answer to microbiological spoilage. Rather, an integrated approach of physical methods, allied with the use of high activity, low toxicity biocides must be adopted.

2. MICROBIOLOGICAL GROWTH IN AND ON WATER–BASED FORMULATIONS

Microorganisms – bacteria, moulds and yeasts (collectively known as fungi) and algae are minute living entities too small to be seen with the unaided eye. They have very simple requirements for growth (Table 1).

These requirements are often met during the manufacture or on storage and use of a very wide range of modern water based products, including paints.

Hence, microorganisms readily become established and the effects of their metabolism soon become evident. During production and on storage, bacteria, moulds and yeasts can cause viscosity loss, gassing, malodour, pH drift, visible surface growth, loss of corrosion inhibition and other undesirable effects. The dry film of products, especially paints, may become infected with fungi and algae, resulting in disfigurement and breakdown of the coating followed by decay or corrosion of the underlying substrate.

Many organisms are able to produce enzymes such as cellulases and amylases which attack cellulosic and starch based products. The former are effective at concentrations as low as 10^{-5} enzyme units per ml in breaking down long chain cellulosic molecules into

Table 1 Growth Requirements of Microorganisms

Requirement	Bacteria	Moulds	Yeasts	Algae
Light	X	X	X	✓
Ideal pH	Slightly alkaline	Slightly acidic	Slightly acidic	Neutral
Ideal temperature	25–40° C	20–35° C	20–35° C	15–30° C
Nutrients	C, H, N sources	C, H, N sources	C, H, N sources	CO_2
Trace elements	✓	✓	✓	✓
Oxygen	O_2 or inorganic, eg. SO_4, NO_3	O_2	O_2	O_2
Water	liquid or vapour	liquid or vapour	liquid or vapour	liquid or vapour

shorter chain oligomers, thereby destroying their viscosity regulating ability[1] and causing dramatic viscosity loss.

The shorter chain length units may then be further degraded to glucose which in turn can be fermented yielding acids and carbon dioxide.

Such microbial acid production can cause a downward drift in pH of as much as 3 units and gas production, especially in the final container, can cause swelling and "lid popping".

Other by–products such as butyric acid and 1,6–Dimethyl–3–methoxypyrazine[2] may impart a rancid or musty odour whilst certain anaerobic organisms, for example the sulphate reducing bacterium *Desulphovibrio desulphuricans*, are able to reduce inorganic sulphate to sulphide liberating hydrogen sulphide leading to a typical "rotten egg smell".

Some pigment producing organisms such as *Serratia sp.* and yeasts such as *Rhodotorula rubra* and *Sporobolomyces roseus* can cause a general discolouration of the product. In addition, the action of sulphate reducing bacteria can result in the formation of metal sulphides thereby causing blackening where gaseous oxygen concentrations are lowest.

Some formulations are particularly prone to the surface growth of organisms, principally moulds, leading to unsightly disfigurement and infection by other deteriogens.

Corrosion can occur as a result of microbial degradation of corrosion inhibitors such as nitrites and alkanolamines. In addition, growth of sulphate reducing bacteria especially in so–called "biofilms" on metal surfaces can, by the process of cathodic depolarisation, cause pitting corrosion and perforation of even stainless steel pipework and tanks[3].

Although virtually any bacterial species can be isolated from water based formulations, a restricted number have been shown to actually grow and cause degradation. For example, studies by five workers[4-8] implicated the species listed by the author[9] in Table 2 as causing degradation of various types of water–based paints.

Table 2   Bacterial Species Isolated from Contaminated Paints by Five Workers

Aerobacter sp.	*Escherichia sp.*
Aerococcus sp.	*Flavobacterium sp.*
Achromobacter sp.	*Hafnia sp.*
Actinomycetes sp.	*Klebsiella sp.*
Alcaligenes sp.	*Micrococcus sp.*
Bacillus sp.	*Proteus sp.*
Citrobacter sp.	*Pseudomonas sp.*
Enterobacter sp.	*Serratia sp.*

Coatings applied to surfaces are not generally degraded by bacterial action. However, in the presence of sufficient moisture and/or light, colonisation by other microorganisms can occur.

Surface growth of moulds, yeasts and algae has been recognised for almost 50 years as the major cause of disfigurement and deterioration of interior and exterior coatings.

A wide range of fungal species have been implicated ranging from Goll and Coffey's 1948[10] report of *Aureobasidium pullulans* being one of the principal fungi involved to Heaton's finding[11] in 1989 that the majority of fungi growing on the walls of a malting kiln were *Penicillia*. The author[12] isolated the species listed in Table 3 from the painted walls of a restaurant's food storage room.

Table 3   Fungi Isolated from a Restaurant Food Store

Aspergillus niger	*Penicillium notatum*
Aspergillus oryzae	*Penicillium purpurogenum*
Aureobasidium pullulans	*Phoma violacea*
Cladosporium sp. 1	*Rhodotorula sp.*
Cladosporium sp. 2	*Saccharomyces sp.*
Paecilomyces variotii	*Sporobolomyces sp.*
Penicillium sp. 1	*Ulocladium sp.*
Penicillium sp. 2	

Fungal growth is most common on interior surfaces especially where conditions are particularly humid such as in bathrooms, shower areas, kitchens and cellars. Airborne spores on germination produce an often invisible mycelium. This, once sporulation has commenced, causes a gradual discolouration of the surface, usually grey/black although, under particularly damp conditions, pink yeasts may predominate producing slimy, coloured growths.

At this stage penetration of the film by fungal hyphae can occur as a result of enzymic breakdown, especially of residual cellulosic thickening agents, by production of organic acids or by simple mechanical disruption. Once this has taken place, cracking, loss of adhesion and blistering may result[13] leading to decay or corrosion of the underlying substrate.

Most growth on exterior painted surfaces is of algae. Uncoated stonework, bricks and paving can be colonised and, although direct deterioration by the organisms is not common, their capacity to hold moisture may result in significant damage.

Algae are simple photosynthetic organisms which obtain their energy from water, CO_2, light and a few trace elements. As a result, their growth in temperate climates[14] is profuse but is even more prolific in the tropics[15].

As with fungi their initial effect is one of disfigurement resulting in green, orange or even black discolouration, depending on the type of algae, the age of the growth and the substrate infected.

Unlike fungi, algae have not been shown to produce significant amounts of corrosive metabolites or degradative enzymes, although Degelius[16], Greathouse and Wessel[17] and Whitely[18] recognised that some species are able to produce organic acids. However, moisture retention on the surface, as a result of algal biofilm development, can cause degradation especially if diurnal temperature variation results in freeze/thaw induced cracking of the film.

Secondary invasion by deteriogenic fungi and by higher plants can then take place and, as with fungi, decay or corrosion of the underlying timber or metal substrate may then occur.

As with fungi, a very wide range of species have been shown to grow on painted surfaces, although a study by Wee and Lee[19] in Singapore (Table 4) named just five species as being the most common.

Table 4   Frequency Occurrences (%) of Algae from a
         Total of 103 Wall Samples in Singapore

Trentepohlia odorata	66	*Schizothrix friesii*	9
Anacystis montana	57	*Oscillatoria lutea*	6
Anacystis thermale	54	*Schizothrix rubella*	6
Chlorococcum sp.	53	*Chlorella sp.*	6
Scytonema hofmanii	34	*Hormidium sp.*	1
Calothrix parietina	19	*Cylindrocapsa sp.*	1
Schizothrix calcicola	12	*Navicula sp.*	1

This resulted in *Trentepohlia odorata* being adopted as the test organism for the Singapore Standard algicidal paint test SS345 Appendix B[20], recently criticised by Downey et al[21].

3. SOURCES OF CONTAMINATION

Microbial contamination especially of products "in the can" may originate from a number of sources (Table 5).

Table 5   Sources of In-Can Microbial Contamination

> Air
> Water
> – make up
> – wash water
> Raw materials
> – powders
> – liquids
> Poor plant hygiene
> Final containers

Air, especially in factory conditions, will contain a wide range of microorganisms and this is particularly the case in dusty environments, common where starches, fillers, talcs, etc. are used.

Water may be taken from boreholes or even rivers, in which case contamination, if untreated, will be high or from the mains supply which can contain non–pathogenic organisms such as *Pseudomonas species*, albeit in low numbers, but which are able to infect many formulations.

Powdered raw materials including fillers/extenders, starches and pigments, especially those which originate from natural sources, will frequently be contaminated with the spores of bacteria and fungi. Once in an aqueous environment these will germinate, grow and cause deterioration.

Many liquid raw materials, especially pigment dispersions, china clay slurries, defoamers and polymer emulsions are themselves susceptible to microbial degradation and, unless manufactured under good conditions and with biocidal protection, can also introduce contamination.

Once applied to a surface, or in the dry state, products and coatings may be infected by fungi or algae principally as a result of spores or microbial propagules being present in the atmosphere either floating freely on air currents or carried on the surface of wind blown dust.

4. PREVENTION AND PROTECTION

4.1 In the wet state

To mitigate against the effects of microbial contamination it is necessary to adopt an integrated approach to control of microorganisms. The use of a biocide alone, however active, is not the complete solution as many manufacturers have found!

The microbiological quality of water, especially any which is recirculated, should be checked. Particular problems may occur if, as for example in the manufacture of some polymer emulsions, deionised water is used. Contamination should be dealt with by the addition of a suitable biocide such as chlorine or by the use of a U.V. steriliser which may be helpful in some instances. Particular attention should be paid to storage tanks since these may become readily contaminated.

The introduction of ISO 9000 quality systems has led to greater attention being paid to incoming raw material quality. Powdered products showing discolouration, abnormal consistency or obvious microbial growth should not be used until thoroughly checked by a competent laboratory. Infection of liquid raw materials is usually even more noticeable and contamination with enzymes may still be a problem even if microorganisms are no longer present.

The need for good plant hygiene cannot be too strongly stressed and this is even more important with the trend away from heavy metal and phenolic containing biocides, such as those based on mercury, tin and pentachlorophenol compounds, to more environmentally acceptable but less persistent types.

For a wide range of water–based formulations a Plant Hygiene Checklist (Table 6) should be followed as far as possible.

Table 6   Plant Hygiene Checklist

* Treat water supply
* Add biocide as first raw material
* Protect stock thickeners with biocide
* Do not allow surface pooling of condensation
* Treat water overlayers with biocide
* Clean down frequently and thoroughly
* Add biocide to residual wash water
* Use a biocidal wash
* Avoid long pipework runs, dead spots, sharp bends
* Keep flexible hoses clean/dry
* Pay attention to filling machinery
* Keep empty containers and lids clean/dry
* Be aware of problems with plastic containers such as
  electrostatic attraction, residual mould release agents
  and excessive condensation
* Keep the factory as clean as possible

Essentially a common sense approach should prevail. If a plant is kept as physically clean as possible, if attention is paid to pipework, hoses, filling lines, filters, mixing vessels, any intermediate storage tanks and final containers, many plant related problems can be avoided.

Such procedures used in conjunction with a broad microbiological spectrum biocide will enable long, trouble–free production runs and consistently good storage stability.

4.2   On the dry film

A similar strategy can be employed for prevention of growth on product surfaces. Measures should be taken to reduce condensation in buildings by modifying their design and improving ventilation, extraction and, where necessary, heating. Allied with thorough cleaning, the use of a biocidal wash and coatings containing high activity fungicides this will result in long term protection from fungal infection[22].

Algal growth occurring on exterior surfaces may be treated following many of the same principles. Elimination of factors which allow a surface to remain damp, e.g. leaking gutters and downpipes, poor building design and overhanging foliage, followed by thorough cleaning, use of a biocidal wash and recoating with an algicidal paint will, again, give prolonged trouble–free performance.

4.3   Biocides

Biocides are regarded by many manufacturers as a "necessary evil". However, with the increasing trend away from solvent based to waterborne formulations, use of, and a greater awareness of the importance of these products is increasing.

Quite often insufficient care is taken to select the correct biocide for a specific application and, since use levels, typically 0.05 to 0.2% for wet state preservation and 0.25 to 2.0% for dry film biocides, are usually relatively low their importance is not sufficiently appreciated until a problem occurs.

The perfect biocide for wet state and dry film preservation will have a number of common and different required properties and those necessary for paints are shown in Table 7.

As can be seen, the main differences are the required microbiological activity spectrum, for wet state against bacteria and fungi, for dry film against fungi and algae and the need for low water solubility/leachability in the dry film application.

Table 7 Properties of the Ideal Paint Biocide

For Wet State Protection	For Dry Film Protection

For Wet State Protection:
* High activity against bacteria/fungi
* pH stable
* Temperature stable
* Water soluble at use concentration
* UV stability in paint not important
* High compatibility
* No effect on colour
* No effect on rheology
* Acceptable toxicity/ecotoxicity
* Cost effective

For Dry Film Protection:
* High activity against fungi/algae
* pH stable
* Temperature stable
* Low water solubility
* UV stability in paint important
* High compatibility
* No effect on colour
* No effect on rheology
* Acceptable toxicity/ecotoxicity
* Cost effective

These differences are important because they make it extremely difficult to develop a single biocide for both applications.

5. BIOCIDES PAST AND PRESENT

A wide range of different biocide types have been used for the preservation of aqueous based formulations during the last 50 years.

With the decline in the use of heavy metal containing biocides such as those based on mercury or tin, no single chemical compound has been shown to give all the necessary properties for both applications.

However, one group of compounds, those based on the five membered heterocyclic isothiazolinone ring, provide a large number of examples of biocides, several of which have been successfully developed on a commercial basis
(Table 8).

BIT, MTIT, CIT and MIT are useful only as wet state preservatives lacking several of the necessary properties to give any appreciable residual dry film protection.

On the other hand OIT performs satisfactorily as a film preservative whilst lacking the necessary spectrum of antibacterial activity and being deficient in several other essential properties of an in-can biocide.

Amongst the wet state isothiazolinone biocides CIT has by far the most potent activity in terms of concentration required to give protection against both bacterial and fungal growth. Typical use concentrations are in the range 10 to 15 ppm of active CIT. In comparison more than 100 ppm of active BIT is often required to give similar antimicrobial activity.

Modern biocide technology has, therefore, concentrated on optimising the activity of CIT.

Table 8   Some Isothiazolinone Biocides

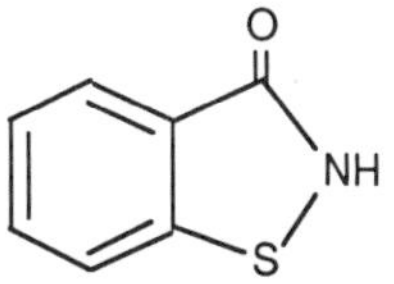

1,2 - Benzisothiazolin
-3-one (BIT)

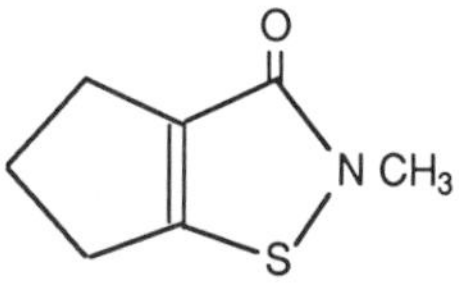

2-Methyl-4, 5-trimethylene-
4-isothiazolin-3-one (MTIT)

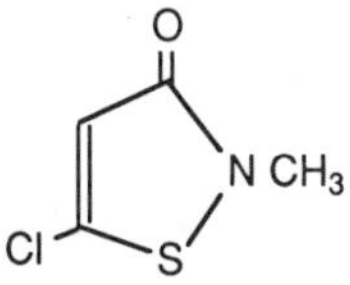

5-chloro-2-methyl-4
-isothiazolin -3-one
(CIT)

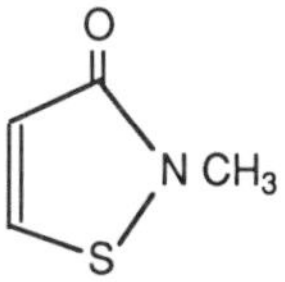

2-methyl-4
-isothiazolin-3-one
(MIT)

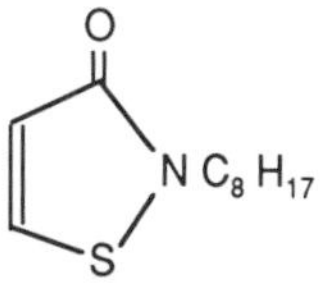

2-n-Octyl-4
-isothiazolin-3-one
(OIT)

5.1   Chloromethyl and Methyl Isothiazolinones

The manufacturing route for CIT results in production of the non chlorinated derivative 2–methyl–4–isothiazolin–3–one (MIT) as a minor component. Normally the two are present in the approximate ratio 3 : 1 CIT : MIT. MIT, whilst being intrinsically more stable than CIT has far lower antimicrobial activity.

The production of CIT/MIT is achieved as follows[23]:–

The methyl ester of acrylic acid is reacted with hydrogen sulphide to form methyl–3–mercaptopropionate which by reaction with methylamine gives the propionamide derivative. With the use of an oxidising agent this can be converted to N,N'–dimethyl–3–3'–dithiopropionamide and this or its monomer is cyclised by means of a dehydrogenation/halogenation agent to the hydrochloride salts of CIT and MIT (Figure 1).

After separation the CIT/MIT blend is then able to be formulated in a number of different ways.

Nucleophilic attack makes CIT in particular readily degradable in the environment[24] and metal salts of nitrates and nitrites are commonly used to stabilize the molecules[25]. One major commercial product contains 14.5% total CIT/MIT and 15% Mg $(NO_3)_2$[26], although the salts of other polyvalent metals, e.g. copper, are also often employed.

Figure 1   Production of Chloromethyl and Methyl Isothiazolinones

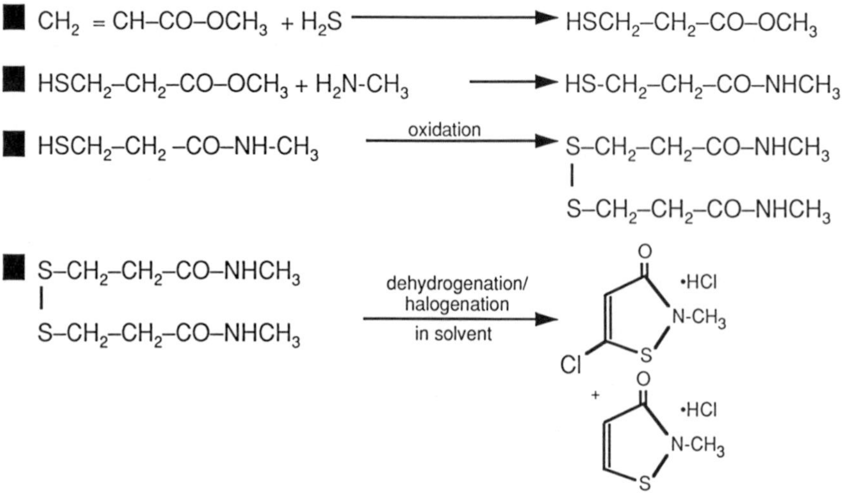

■ $CH_2 = CH-CO-OCH_3 + H_2S \longrightarrow HSCH_2-CH_2-CO-OCH_3$

■ $HSCH_2-CH_2-CO-OCH_3 + H_2N-CH_3 \longrightarrow HS-CH_2-CH_2-CO-NHCH_3$

■ $HSCH_2-CH_2-CO-NH-CH_3 \xrightarrow{\text{oxidation}} S-CH_2-CH_2-CO-NHCH_3$
$\quad\quad\quad\quad\quad\quad\quad\quad\quad\quad\quad\quad\quad\quad\quad\quad\quad | $
$\quad\quad\quad\quad\quad\quad\quad\quad\quad\quad\quad\quad\quad\quad\quad\quad S-CH_2-CH_2-CO-NHCH_3$

■ $S-CH_2-CH_2-CO-NHCH_3$
$\;|$
$S-CH_2-CH_2-CO-NHCH_3$

dehydrogenation/
halogenation
in solvent

Riha et al[27] investigated Rossmoore's view that not only does Cu^{2+} prevent nucleophiles from degrading CIT/MIT but, since all vital molecules in the bacterial cell are nucleophilic, Cu^{2+} might also interact with cellular components in the same way and by blocking non–vital nucleophiles, Cu^{2+} would save the CIT/MIT for the lethal targets. Riha et al's work involving sequential treatment of a bacterial population with Cu^{2+} and CIT/MIT, indicated that pretreatment with Cu^{2+} enhanced the activity of the isothiazolinones.

Hence the polyvalent metals probably not only stabilize the isothiazolinones but also enhance their activity.

Without stabilization 5–chloro–2–methyl–4–isothiazolin–3–one follows a degradation pathway similar to its route of breakdown in the environment. (Figure 2)[24]

Ring opening and production of elemental sulphur results in formation of methylamine hydrochloride and malonic acid which, with the release of carbon dioxide yields acetic and formic acids.

5.2   By–Products of Isothiazolinone Manufacture

One of the unwanted by–products likely to be formed during the manufacture of the CIT/MIT blend is 4,5–Dichloro–2–methyl–4–isothiazolin–3–one (DCIT – Figure 3).

Figure 2   Degradation Pathway of 5–chloro–2–methyl–4–
isothiazolin–3–one and 2–methyl–4–isothiazolin–3–one

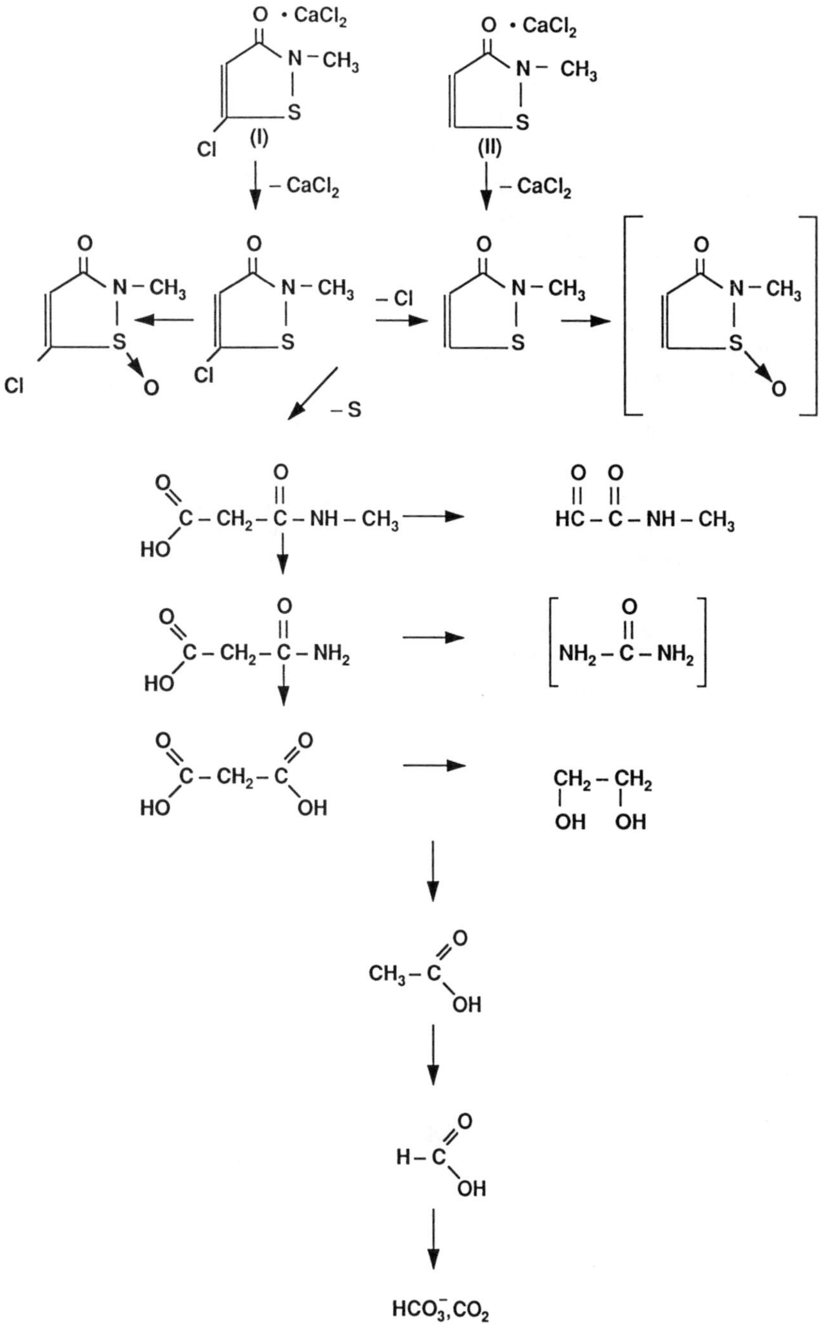

Figure 3   4,5–Dichloro–2–methyl–4–isothiazolin–3–one

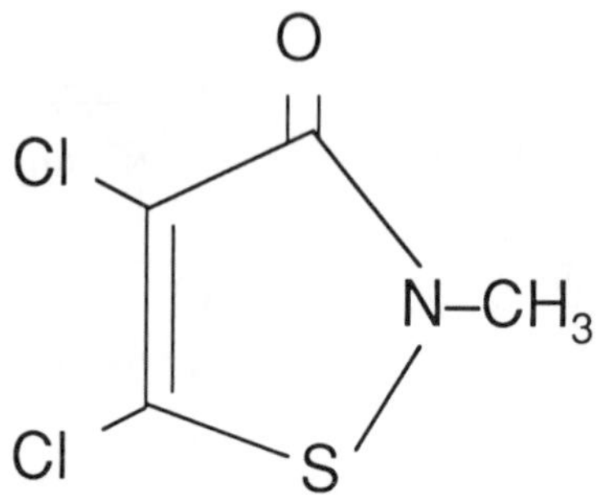

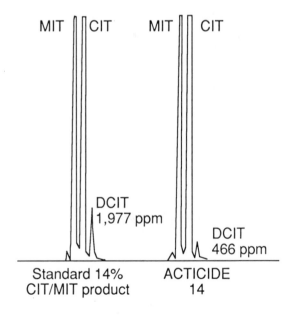

This has been shown to be a particularly potent skin sensitizer[28] and all animals in the study which became sensitized to DCIT also reacted to CIT when rechallenged.

By taking great care during the manufacturing process it is relatively easy to produce the CIT/MIT blend containing less than 500 ppm of DCIT rather than the almost 2,000 ppm which has been routinely monitored in other commercially available products and in the near future this will be reduced to less than 100 ppm. Comparative analyses by HPLC of a standard 14% CIT/MIT product and Thor Chemical's own 14% blend – ACTICIDE 14 indicated 1,977 ppm of DCIT in the former product with only 466 ppm being detected in the latter (Figure 4).

Figure 4   Comparative 4,5–Dichloro–2–methyl–4–isothiazolin–3–one
           contents of Two Biocides by HPLC

5.3 Formulation of CIT/MIT Biocides

It is relatively simple, given good quality base 14% CIT/MIT, to produce stable formulations, amongst which the most common are 1.5% total isothiazolinone products stablized either with copper or magnesium salts, or a combination of these.

One consequence of the use of polyvalent metal ion based stabilization systems is their incompatibility with certain formulations and their effect on the sheer stability of polymer emulsions. This can be overcome by the use of a polyvalent metal and nitrate ion–free stabilization agent such as that developed by Thor Chemicals for their latex grade biocide ACTICIDE LG.

In addition formulation with formaldehyde and formaldehyde adducts such as N–methylol–chloroacetamide and 2–Bromo–2–nitro–1,3–propanediol can result in products with improved properties compared with a straight 1.5% blend.

Rossmoore[26] considered the similarities in reactivities of isothiazolinones (IT) and formaldehyde with nucleophiles and enhancement by Cu^{2+} and examined the possibility that IT and formaldehyde might interact favourably.

Referring to the extensive listing of industrial biocides by Allsopp and Allsopp[29] he studied the interactions between copper, formaldehyde, formaldehyde adducts and isothiazolinone and concluded that synergistic activity between these could be demonstrated.

5.4 2–n–Octyl–4–isothiazolin–3–one

2–n–Octyl–4–isothiazolin–3–one (OIT) is prepared according to the following five stage reaction.

Following production of methyl–3–mercaptopropionate from the methyl ester of acrylic acid and hydrogen sulphide, reaction with 1–octylamine forms 1–octyl–3–mercaptopropionamide which by oxidation gives N,N'–dioctyl–3,3'–dithiopropionamide. Dehydrogenation and halogenation produces the hydrochloride salt of OIT from which the free base is prepared (Figure 5).

The final product sold by Thor Chemicals as ACTICIDE 45 is normally a 45% solution in 1,2–Propanediol and has many applications including fungicidal protection of paints, anti–sapstain protection of timber as well as for antifungal treatment of wood stains, leather, mastics, inks, sealants, adhesives, paper and pulp, metalworking fluids, etc.

However, as with the chloromethyl and methyl derivatives (CIT and MIT) OIT may be inactivated under certain conditions by the presence of ammonia, primary and secondary amines, strong reducing or oxidizing agents such as hypochlorites, bisulphites, metabisulphites and H_2S. Therefore, its use may be restricted in certain formulations.

Additionally its water solubility (480 ppm) is rather high for a product intended to give long term antifungal protection to applied surface coatings. Although it may become bound into the structure of certain coatings' formulations, its susceptibility to leaching can be problematic in high humidity/high rainfall areas.

Figure 5   Production of 2–n–Octyl–4–isothiazolin–3–one

■ $CH_2 = CH-CO-OCH_3 + H_2S \longrightarrow HSCH_2-CH_2-CO-OCH_3$

■ $HSCH_2-CH_2-CO-OCH_3 + H_2NC_8H_{17-n} \longrightarrow HSCH_2-CH_2-CO-NHC_8H_{17-n}$

■ $HSCH_2-CH_2-CO-NHC_8H_{7-n} \xrightarrow{\text{oxidation}}$

$S-CH_2-CH_2-CO-NHC_8H_{17-n}$
$|$
$S-CH_2-CH_2-CO-NHC_8H_{17-n}$

■ $\begin{array}{l} S-CH_2-CH_2-CO-NHC_8H_{17-n} \\ | \\ S-CH_2-CH_2-CO-NHC_8H_{17-n} \end{array} \xrightarrow[\text{agent in solvent}]{\substack{\text{dehydrogenation/} \\ \text{halogenation}}}$

(structure: isothiazolinone ·HCl with N C_8H_{17-n})

■ (structure: isothiazolinone ·HCl with N C_8H_{17-n}) $\xrightarrow[- HCl]{H_2O}$ (structure: isothiazolinone with N C_8H_{17-n})

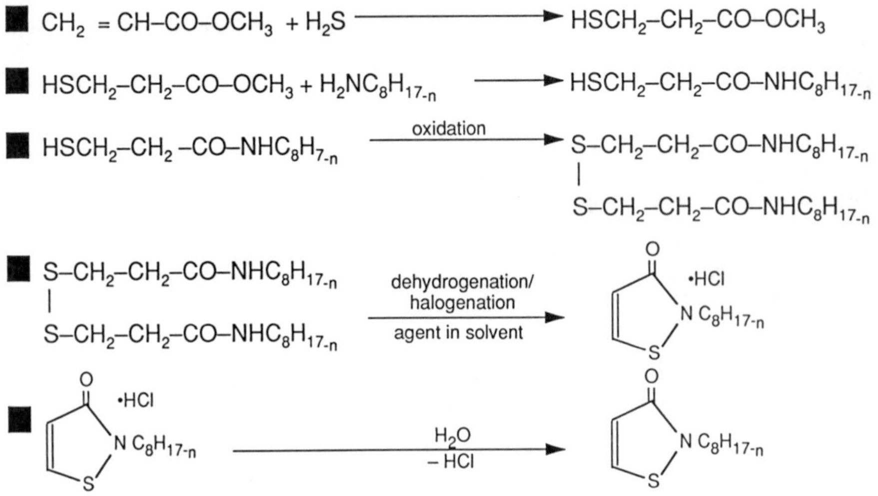

Also, although in minimum Inhibitory Concentration Tests (MICs) it inhibits Chlorophycae (Green Algae) and Cyanophycae (Blue Green Algae) at between 0.5 and 5.0 ppm, in practical tests such as that described by Gillatt[30], it has little paint film algicide activity after 350 hours weathering to BS3900 Part F3.

5.5 Formulations of 2–n–Octyl–4–isothiazolin–3–one

Despite its weaknesses OIT is an extremely useful compound, especially when formulated with other active agents to improve its stability and microbiological activity spectrum.

Stability can be enhanced by blending with inorganic compounds such as zinc oxide and microbiological performance, especially to increase antialgal activity in exterior coatings, can be boosted by careful formulation with a specific algicide. Additional antifungal agents are also often blended into the formulation, producing a stable, high activity fungicide/algicide suitable for use in the most inhospitable environments.

6. CONCLUSION

Waterborne coatings present a challenge to the Biocide manufacturer as well as the paint producer.

However, by carefully controlling environmental factors which influence microbial growth in and on waterborne paints including factory hygiene and practice, good raw material quality control, moisture on and soiling of surfaces, significant improvements in product quality can be achieved.

Allied with the use of broad spectrum biocides such as those formulated around isothiazolinone derivatives, long lasting microbiologically trouble free performance can be attained.

REFERENCES

1. Gillatt, J.W., *Surface Coatings International*, 20, 387–392, 1992.

2. Mottram, D.S. and Patterson, R.L.S., *Chemistry and Industry*, 18th June 1984, 448–449.

3. Hardy, J.A., *"Microbial Problems and Corrosion in Oil and Oil Product Storage"*, Institute of Petroleum, London, UK., 5th October 1983.

4. Huddard, G., *"In–can Preservatives for Emulsion Paints"*, internal company publication, February 1983.

5. Jakubowski, J.A., Gyuris, J. and Simpson, S.L. (1983), Microbiology of Modern Coatings Systems, *Journal of Coatings Technology*, 55, (707), 49, 1983.

6. Miller, W.G., *Journal of the Oil and Colour Chemists' Association*, 56 (7), 307, 1973.

7. Opperman, R.A. and Goll, M., *Journal of Coatings Technology*, 56 (712), 51, 1984.

8. Woods, W.B., *Journal of Waterborne Coatings*, November 1982, 2.

9. Gillatt, J.W., *Surface Coatings International*, 74, 9, 324–328, 1991.

10. Goll, M and Coffey, G., *Paint Oil and Chemical Review III*, 16, 14, 1948.

11. Heaton, P.E., Butler, G.M. and Callow, M.E., *International Biodeterioration*, 16, 1, 1–9, 1990.

12. Gillatt, J.W., *"Airborne Deteriogens and Pathogens:– Proceedings of the Spring Meeting of the Biodeterioration Society, Occasional Publication 6"*, 183–189, 1989.

13. Holbrow, G.L. *Notes to Industry No. 14*, Paint Research Association, UK, 1984.

14. Grant, C., *International Biodeterioration Bulletin*, 18, 3, 57–65, 1982.

15. Chua, N.H., Kwok, S.W., Tan, K.K., Teo, S.P. and Wong, H.A., *Journal of the Singapore Institute of Architects*, 51, 13–15, 1972.

16. Degelius, G., *"Chemie im Dienst der Archaelogie Bautechnik und Denkmalpflege"*, 156–163, 1962.

17.   Greathouse, G.A. and Wessel, C.J., *"Deterioration of Materials"*, Reinhold Publishing Corpn., New York, USA., 1954.

18.   Whiteley, P., *Journal of the Oil and Colour Chemists' Association*, 56, 382–287, 1973.

19.   Wee, Y.C. and Lee, K.B., *International Biodeterioration Bulletin*, 16, 113–117, 1980.

20.   Singapore Institute of Standards and Industrial Research, *"Specification for Algae Resistant Emulsion Paint for Decorative Purposes 55345, Appendix B"*, 1990.

21.   Downey, A. and Frazier, V.S., Paper presented at "The 9th International Biodeterioration and Biodegradation Symposium", University of Leeds, UK, August 1993.

22.   Building Research Establishment, *"Defect Action Sheet DAS 16"*, Housing Defects Prevention Unit, Building Research Establishment, Garson, Watford, UK, 1983.

23.   Lewis, S.N., Miller, G.A., Hausman, M. and Szamborski, E.C., *Journal of Heterocyclic Chemistry*, 8, 571–580, 1971.

24.   Krzeminski, S.F., Brackett, C.K., Fisher, J.D. and Spinnler J.F *Journal of Agriculture and Food Chemistry*, 23, 6, 1068–1075, 1975.

25.   Miller, G.A. and Weiler, E.D., *U.S. Patent 4067878*, 1978.

26.   Rossmoore, H.W., *International Biodeterioration, Special Issue – Biocides*, 26, 2–4, 225–235, 1990.

27.   Riha, V.F., Sondossi, M. and Rossmoore, H.W., *International Biodeterioration*, 26, 1, 51–61, 1990.

28.   Bruze, M., Gruvberger, B. and Persson, K. (1987). Contact allergy to a contaminant in Kathon CG in the guinea pig. *Dermatosen 5*, 165–168, 1987.

29.   Allsopp, C. and Allsopp, D., *International Biodeterioration Bulletin*, 19, 99–146, 1983.

30.   Gillatt, J.W., *Surface Coatings International*, 74, 6, 197–203, 1991.

Synthetic Clay Rheology Modifiers for Water Based Coatings

P. K. Jenness

LAPORTE ABSORBENTS EUROPE, PO BOX 2, MOORFIELD ROAD, WIDNES, CHESHIRE
WA8 0JU, UK

1. INTRODUCTION

Clays such as bentonite, hectorite and attapulgite are used as rheology modifiers in a wide range of formulated products including coatings, household cleansers and personal care products. As naturally occurring minerals these clays vary considerably in appearance, composition and in potential end-use. Even with a single clay source variations will occur through the deposit which will result in variation in finished product performance. Whilst sources of hectorite and attapulgite clay tend to be limited to the USA commercially viable sources of the much more abundant mineral bentonite (or more correctly sodium montmorillonite) are found throughout the world. The performance of these clays may be improved by removal of impurities such as quartz and calcite and chemical treatments such as activation with sodium are often carried out. It is possible to mimic the processes which occurred to produce natural clays and manufacture synthetic analogues of some types of clay mineral. Laporte manufacture a range of synthetic hectorites under the trade name Laponite which resemble the natural mineral hectorite in both structure and composition. The synthesis process (Figure1) involves combining solutions of sodium carbonate, magnesium sulphate, lithium carbonate and sodium silicate at carefully controlled rates and temperatures. This produces an amorphous precipitate which is partially crystallized by hydrothermal treatment. The resulting product is then filtered, washed, dried and milled to a fine powder.

Synthetic hectorite has a number of distinct advantages when compared with natural clays. Its efficiency as a thickener is many times higher and it can be dispersed rapidly

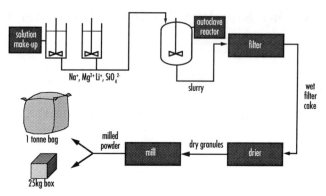

Figure 1 *Production of synthetic hectorite*

without the need for high shear mixing or elevated temperature. Sol forming grades can be produced which disperse to give water-thin concentrates that will gel on addition to a formulated product. Additionally, as a synthetic material it is both chemically and mineralogically pure and by closely controlling the manufacturing process it can be produced to a high level of consistency.

The benefits of synthetic hectorite over its naturally occurring countertypes can be explained by an examination of its chemistry and structure, the crystal morphology and the mechanisms involved in dispersion of the clay to produce gels.

2. CHEMISTRY AND STRUCTURE

Hectorite is a layered hydrous magnesium silicate which occurs as thin, platelet-like crystals. The chemical composition of clays like hectorites and montmorillonite can be simplified into an idealized empirical formula or unit cell based upon eight silicon atoms. This unit cell will be repeated many thousands of times to give a single primary particle or crystal of the clay.

Typically a naturally occurring Wyoming sodium montmorillonite has the idealized empirical formula (1).

$$Na_{0.5}^{0.5+} [(Si_8 Al_{3.1} Fe_{0.4}^{III} Mg_{0.5})O_{20}(OH)_4]^{0.5-}$$

(1)

For Californian hectorite the idealized empirical formula is shown in (2).

$$Na_{0.62}^{0.62+} [(Si_8 Al_{0.04} Mg_{5.3} Li_{0.66})O_{20} F_2 (OH)_2]^{0.62-}$$

(2)

The formula for synthetic hectorite is shown in (3)

$$Na_{0.7}^{0.7+} [(Si_8Mg_{5.5}Li_{0.3})O_{20}(OH)_4]^{0.7-}$$

(3)

The idealized structure of hectorite is shown in Figure 2.

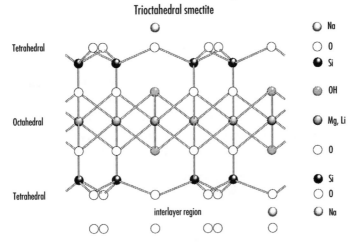

Figure 2. *Idealized structure of a unit cell of hectorite*

The unit cell consists of six octahedral magnesium ions sandwiched between two layers of four tetrahedrally oriented silicon atoms. In synthetic hectorite these groups are balanced by twenty oxygen atoms and four hydroxyl groups. In natural hectorite the balance is made up by twenty oxygen atoms, two hydroxyl groups and two fluorine atoms.

The idealized structure depicted in Figure 2 has a neutral charge with six magnesium ions in the octahedral layer giving a positive charge of twelve. In practice, however, in synthetic hectorite, magnesium ions are substituted by lithium ions and some of the octahedral positions are empty giving an octahedral layer positive charge of:

Mg^{2+}	5.5 x 2	=	11.0
Li^+	0.3 x 1	=	0.3
	Total		11.3

Thus the unit cell has a charge deficiency of 12 - 11.3 = 0.7. This negative charge is neutralized by the adsorption of sodium ions onto the surfaces of the synthetic hectorite crystal. During drying the crystals or platelets arrange themselves into stacks which are held together electrostatically by the sharing of sodium ions in the interlayer region between the crystals.

By making modifications to the composition of the reaction mixture it is possible to increase or decrease the anionic charge on a synthetic hectorite crystal. As the charge is increased the speed of dispersion is greatly reduced and subsequent gel formation is much slower. Alternatively, if the anionic charge is reduced significantly then gel formation can become so rapid that gel coated clumps of dry powder are formed when the synthetic clay is added to water.

Natural hectorite has a similar arrangement but has a higher level of magnesium substitution by lithium. A small amount of aluminium is also present. The montmorillonite (bentonite) structure is analogous to that of hectorite, however, the idealized octahedral layer is composed of four aluminium ions, with substitution by magnesium providing the charge deficiency. Depending upon the source of the montmorillonite this can be neutralized by sodium, calcium and in some cases magnesium ions.

Although the chemical composition and structural arrangement of synthetic and natural clays are similar a significant difference is seen in the very much smaller size of the primary crystal or platelet of the synthetic hectorite (Figure 3).

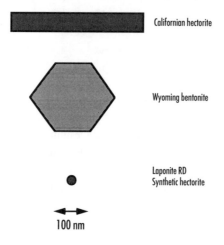

Figure 3 *Comparison of clay primary particle size*

The primary particle of all have similar thickness, approximately 1nm, but the natural clays are an order of magnitude larger in both breadth and width. Synthetic hectorite forms colourless completely transparent dispersions in water because of its high purity and because the primary particles are small in size (25nm in diameter) compared with the wavelengths of visible light (400-700nm). Natural clays invariably contain mineral and transition metal impurities and dispersions can be obtained showing a wide range of colours from cream through to pinks, greens and browns. Dispersions of natural hectorites and montmorillonites are opaque because of their relatively larger particle dimensions (typically 200-1000nm).

3. ADDITION OF SYNTHETIC HECTORITE TO WATER

A description of the processes occurring during the dispersion and gel formation of synthetic hectorite has been developed from work carried out by Verway and Overbeek,[1] Van Olphen,[2] and Van Oss et al.[3]

3.1 Dispersion.

The dispersion of synthetic hectorite is depicted graphically in Figure 4.

At room temperature in normal tap water with low speed mixing this process is substantially complete for synthetic hectorite after 10 minutes. Complete hydration of the sodium ions within the much larger platelets of natural clays takes longer as the water molecules must penetrate greater distances into the interlayer. Additionally, traces of

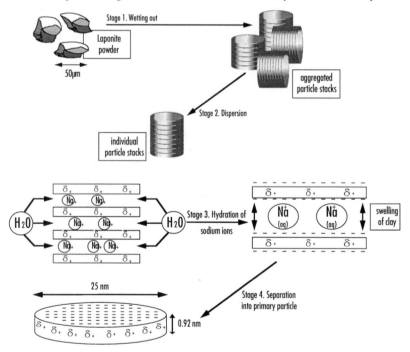

Figure 4 *Dispersion of synthetic hectorite*

calcium ions which partially substitute for sodium ions increase the total energy required for complete hydration. Because of this full dispersion is difficult to achieve and the use of high shear mixing equipment, elevated temperature or chemical dispersing agents are required.

3.2 Gel Formation

Dilute suspensions of synthetic hectorite in water with low electrolyte levels will remain as low viscosity sols of non-interacting individual platelets for long time periods. [4,5] The clay crystal surface has a negative charge of 50-55 meq/100g. The edges of the platelet have small localized positive charges generated by the adsorption of hydroxyl groups where the crystal structure terminates. This positive charge is typically 4-5 meq/100g.

The solvated sodium ions are held in a diffuse region on either side of the clay composite layers. Electrostatic attractions draw the cations back towards the clay surface; osmotic pressure from the bulk of water holds the cations away (Figure 5). This interaction forms an electrical double layer of negative charge on the clay crystal surface and a diffuse cloud of positive charges surrounding it. When two particles approach each other their mutual positive charges repel each other.

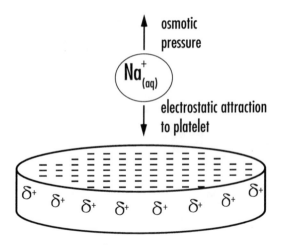

Figure 5 *Dispersed primary particle*

The addition of polar compounds in solution (simple salts, surfactants, wetting agents, coalescing solvents, soluble impurities and additives in pigments etc.) will reduce the osmotic pressure holding the sodium ions away from the clay particle surface, causing the electrical double layer to thin and effectively reducing the net negative charge on the crystal surface. This process allows associations to occur between the positively charged edges and negatively charged faces which result in the classic house of cards structure shown in Figure 6 being formed.

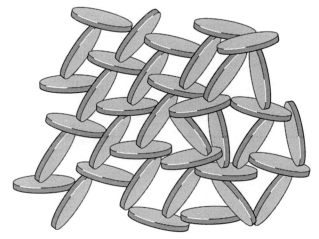

Figure 6 *House of cards structure*

The gel consists of a single flocculated particle held together by weak electrostatic forces. These forces are comparable in strength between synthetic hectorite and natural hectorites and bentonites. However, because the synthetic product is more completely dispersed into its primary particles and those particles are an order of magnitude smaller in size many more edge-face interactions will be produced for the same mass of synthetic clay than for a natural clay. Consequently the synthetic analogue is between three and eight times more efficient as a thickener than the natural hectorites and bentonites.

A number of the features of the rheology of synthetic hectorite support this type of mechanism for gel formation.

3.2.1. Solid particles are held within the 3D gel structure and are not stabilized by viscosity alone. This results in excellent suspension properties for materials of all densities.

3.2.2. The gel structure is readily broken down on application of shear stress. Synthetic hectorite shows a greater degree of shear thinning than other thickeners commonly used in paint including gums, cellulose ethers, metal chelate complexes, associative thickeners and natural clays.[6]

A simple dispersion of Laponite in water at 2% concentration will have a viscosity of 4×10^6P at a shear rate of $10^{-4}s^{-1}$. At a higher shear rate of 10^3s^{-1} this viscosity falls dramatically to only 30 cP (Figure 7).

3.2.3. Similarly, when held under high shear synthetic hectorite dispersions show very little resistance to flow and have low viscosity. Figure 8 shows the effect of adding a low

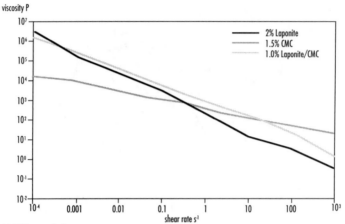

Figure 7 *Effect of shear*

concentration of Laponite RDS synthetic hectorite dispersion to a commercial paint formulated with HEC. A logarithmic increase in viscosity is seen at lower shear rates combined with only a small increase in high shear viscosity.

3.2.4. The gel structure takes time to reform when shear stress is removed as the particles must re-orientate themselves into the house of cards structure.

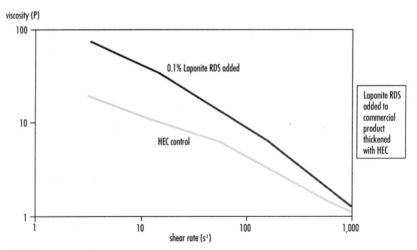

Figure 8 *Effect on paint viscosity*

4. SYNERGY WITH CELLULOSE THICKENERS

Figures 7 and 8 also demonstrate the synergistic increases in viscosity seen when synthetic hectorite is used in combination with cellulosic thickeners. At only 1% concentration a 50:50 blend of Laponite RD and sodium carboxymethyl cellulose has a viscosity higher than would be predicted by observing the viscosity of the two individual thickeners (Figure 7). This synergistic behaviour has been demonstrated by combinations of Laponite RD with other cellulose ethers including HEC (Figure 8), hydrophobically modified HEC, MHEC and MC.[6]

5. SOL FORMING SYNTHETIC HECTORITE

Two general definitions:

> A high viscosity colloidal dispersion is termed a ***gel***.
> A low viscosity colloidal dispersion is termed a ***sol***.

It is possible to modify synthetic hectorite from a gel forming type to a sol forming type by addition of certain compounds, for example tetrasodium pyrophosphate (TSPP). When synthetic hectorite is added to a solution of TSPP it will disperse in the same manner as described in Section 3.1. The pyrophosphate anions will associate with the positively charged edges of the synthetic hectorite crystals making the whole particle negatively charged (Figure 9). This then becomes surrounded by a diffuse cloud of loosely held sodium ions which cause the clay platelets to repel each other and allows the preparation of a colloidal sol concentrate which shows a low Newtonian type viscosity. At a concentration of 11.5% in water Laponite RDS will have viscosity of 10-15 cP. For longer term storage of stock dispersions a working concentration of 8-10% is recommended.

When this dispersion is added into a formulated product such as a paint the peptizing effect of the TSPP is rapidly overcome as the pyrophosphate anions are adsorbed into the matrix of fillers, pigments, binders, surfactants, wetting agents etc. that comprise the paint formulation. As this occurs the clay particles become free to arrange themselves into the house of cards type structure described earlier and the paint will thicken. This unique feature of synthetic hectorite gives the opportunity to develop a viscosity increase at a selected time during formulation. In this way it can be used as a post-additive or corrective thickener. A particularly interesting use of the sol forming property is its ability to stabilize a water based multicolour paint. This is described in detail in Section 7.

Synthetic hectorite/TSPP powder blends have been developed which contain a level of TSPP optimized for maximum sol stability at low viscosity.

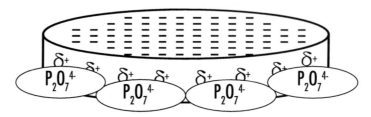

Figure 9 *Individual particles stabilised by peptizer*

6. FORMULATION OF PAINTS WITH SYNTHETIC HECTORITE

As with most thickeners following the correct order of addition of synthetic hectorite to a formulation is crucial in order to obtain optimum performance. The clay should be added to water alone or with a cellulosic co-thickener and allowed to hydrate fully before addition of other components. A commonly made mistake is to add thickening clays at the same time as the pigments and extenders. The presence of components already in solution such as surfactants, dispersants etc. interfere with the dispersion of clays and in some cases prevent it completely. With sol forming synthetic hectorite it is possible to prepare a concentrate in advance which can be used as a stock for addition to a succession of paint batches.

Synthetic hectorite is ideal for developing the type of thixotropic gel paint which is popular in the UK and north European markets. Combination of clays with cellulose ethers such as those described in Section 4 allows the paint formulator to design a rheological profile suited to a particular end application. Synthetic hectorite provides high viscosity at low shear giving good suspension of pigment and long term in-can stability. The high degree of shear thinning down to a low viscosity gives improved application by brush or spray with good levelling after application. The short (non-elastic) texture characteristic of clays helps to reduce roller spatter. Modification of this rheology with a cellulose ether gives the paint more body and increases the rate of recovery of viscosity after shear to prevent sag. Varying the ratios of the two types of thickeners allows the development of a formulation which shows the optimum compromise of long term stability and application properties. Two example formulations are shown below which show widely different rheological properties. The first, a smooth masonry paint, contains a relatively high level of Laponite RD synthetic hectorite and produces a strongly gelled paint which readily shear thins during application. The second, a contract vinyl matt, contains a lower level of Laponite RDS modified synthetic hectorite (2% of a 10% solids dispersion) added as a final stage before the let-down with water. This has a lower gel strength and a creamy

Smooth Masonry Paint

Order of addition	% weight	Volume
Water	15.00	15.00
Natrosol Plus	0.24	0.24
Laponite RD	0.45	0.45
Premix 600 rpm/15 minutes		
Dispex G40	0.25	0.21
CF 107	0.10	0.11
Acticide BX	0.15	0.13
Acticide EP Paste	0.40	0.31
TiONa 535	20.00	4.76
Speswhite clay	3.00	1.11
Snowcal 60	3.00	1.11
Microtalc AT 200	6.00	2.07
Disperse at high shear for 20 minutes		
Add stepwise with slow speed mixing		
Vinamul 3459	28.00	26.17
CF 107	0.10	0.11
Texanol	0.20	0.21
Water	22.86	22.86
pH adjustment (0.91 Ammonia)	0.25	0.27
Total	100.00	75.12

Non-volatile content by weight = 48.09%
Non-volatile content by volume = 30.79%
pH = 8.20 @ 20°C
S.G = 1.33 @ 20°C
PVC = 40%

ICI Rotothinner viscosity =10.5P
ICI Gel Strength = 45 gm/cm
Stormer viscosity = 112 KU

Contract Vinyl Matt

Order of addition	% weight	Volume
Water	18.00	18.00
Natrosol Plus 330	0.35	0.35
Premix 600 rpm/15 minutes		
Dispex G40	0.25	0.22
CF 107	0.05	0.05
Acticide BX	0.15	0.13
TiONa 376	12.00	3.33
Microdol AT Extra	12.00	4.21
Talc 2628	6.00	10.34
Speswhite clay	10.00	3.70
Disperse at high shear for 20 minutes.		
Add stepwise with slow speed mixing		
Harco VV 575	12.00	11.11
CF 107	0.05	0.05
Texanol	1.00	1.05
10% Laponite RDS	2.00	2.00
Water	25.95	25.95
pH adjustment (0.91 Ammonia)	0.20	0.22
Total	100.00	80.71

Non-volatile content by weight = 47.15%
Non-volatile content by volume = 34.53%
pH = 8.6 @ 20°C
S.G = 1.24 @ 20°C
PVC = 79%

ICI Rotothinner viscosity =7P
ICI Gel Strength = 12 gm/cm
Stormer viscosity = 89 KU

texture. Both formulations show good long term storage stability and application properties.

7. PROCEDURE FOR PREPARATION OF WATER BASED MULTICOLOUR PAINT USING LAPONITE RD SYNTHETIC HECTORITE

The following describes a procedure for preparation of a very low VOC water based multicolour paint for airless spray application. It utilizes the unique ability of synthetic hectorite to form concentrated low viscosity colloidal sols.

The formulation involves preparation of a single base paint which is then split into two or more portions which are tinted as required using pigment dispersions. The separate colours are blended together with a stabilized sol dispersion of synthetic hectorite. By following the mixing procedure described, a product is formed where the individual paint particles or flecks are covered in a thin layer of synthetic hectorite gel. This gel coating acts as a physical barrier to prevent the different colours mixing together. The exceptional shear thinning shown by these gels has the additional property of providing a lubricant between the paint particles to prevent their blending during mixing and spray application.

The formulation has proven to show good long term storage and application stability.

It should be emphasized that this procedure has been simplified to illustrate the basic property of producing a stable water based multicolour paint. In practice in developing a commercial multicolour paint it is likely that the secondary or minor colours will be made to a different formulation than the primary or major colour. Modification of the rheology of the secondary colour to give a more elastic texture can allow the production of a paint which has a larger coloured fleck size in the coated film. This may be achieved by replacing part of the Laponite synthetic hectorite and hydroxyethyl cellulose thickening system shown with thickeners which show extended, elastic viscosity in the higher shear range. Polyurethane associatives and some organic gums are suitable. It has, however, been found that complete removal of Laponite from the base paint results in reduced gel strength and allows the colours to mix with each other to give a single coloured paint.

The formulation has been divided into three stages:

Stage 1 *Preparation of base paint*
A highly gelled paint is prepared to the following formulation.

Order of addition	% weight
Water	35.47
Laponite RD	0.42
Mix until Laponite is fully hydrated	
Natrosol 250 MHR	0.42
Mix and add the following	
Dispersant (Tamol 731)	0.66
Wetting agent (Triton CF-10)	0.19
Defoamer (Dispelair CF107)	0.19
Biocide (Acticide AZ)	0.09
Titanium dioxide (RCL 535)	4.73
Calcium carbonate (Queensfill 300)	23.65
Disperse to 5 Hegman, then add	
Latex (Viking 2920)	28.37
Defoamer (Dispelair CF 107)	0.09
Water	5.72
Total	100.0

Stage 2 *Tinting*
The primary and secondary colours should be prepared from the base paint in the ratios required. In practice, the actual proportion of each colour varies widely depending upon manufacturer and aesthetic requirements. For the purpose of showing an example Stage 3 describes a paint composed of

86.8 parts white
6.6 parts blue
6.6 parts red

Multicolour paints covering a much wider colour spectrum are, of course, possible.

Tinting should be carried out by adding water based pigment concentrates to the level required. It is highly recommended that water based pigment pastes are used for maximum long term colour stability.

Stage 3 *Blending of the multicolour product*

In this final stage the separate colours are successively added to a colloidal sol dispersion of Laponite synthetic hectorite which has been stabilized by tetrasodium pyrophosphate. When the paint comes into contact with the synthetic hectorite sol a thin coating of gel is formed on the surface of the paint. This is caused by the interaction between the polar compounds present in the paint with the tetrasodium pyrophosphate.

Slow speed mixing (50 rpm or less) causes the paint to break up into small particles which are separated from each other by the formation of the protective coating of synthetic hectorite gel. This is shown schematically in Figure 10.

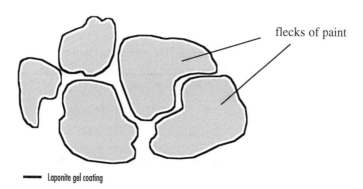

flecks of paint

—— Laponite gel coating

Figure 10 *Paint flecks coated in Laponite gel*

A colloidal styrene acrylic emulsion is added in the final stages to incorporate the free synthetic hectorite firmly into the paint film on drying and give improved scrub resistance.

The formulation is shown below

Order of addition	% weight
Water	41.57
Tetrasodium pyrophosphate	0.04

Mix until TSPP is dissolved

Laponite RD	2.00

Mix until Laponite is fully hydrated, then add

White base paint	32.80
Blue tinted base	2.50
Red tinted base	2.50
Water	10.80
Styrene Acrylic emulsion (Viking 5455)	7.56
Coalescent (Texanol)	0.23
Total	100.0

Notes on application

Before use stir the paint gently. Do not shear. Do not thin. Apply by spray in cross lay method. Use high volume low pressure (HVLP) equipment. The Sata NT HVLP or Binks Mach 1 HVLP spray guns are suitable. Pot pressure should be 10 psi (0.7 bar) and the air pressure at the gun should be 5-10 psi (0.35-0.70 bar) with inlet pressure to the gun being 65-75 psi (4.5-5.2 bar).

As with all speciality coatings the appearance and quality of the finished coatings are highly dependent upon the technique and skill of the operator.

8. SUMMARY

Four key properties of synthetic hectorite recommend its use in a wide range of water based coatings.

■ High gel strength

■ Highly shear thinning

■ Forms versatile low viscosity sol concentrates

■ Synergistic with cellulose ethers

References

1. Verwey E.J.W., Overbeek J. Th. G. : "The Theory of the Stability of Lyophobic Colloids" : Elsevier (Amsterdam) (1948).

3. Van Olphen H : "An Introduction to Clay Colloid Chemistry" 2nd Edition, Wiley (New York) (1977).

3. Van Oss C. J., Giese R.F., Constanzo P. M. : "DLVO and non DLVO Interactions in Hectorite" Clays and Clay Colloids : 38(2), 151 (1990).

4. Ramsey J. D. F. : "Colloidal Properties of Synthetic Hectorite Dispersions: I Rheology" : J. Colloid Interface Science : 109(2), 441 (1986)

5. Avery R. G., Ramsey J. D. F. : "Colloid Properties of Synthetic Hectorite Dispersions : II Light Scattering and Small Angle Neutron Scattering" : J. Colloid Interface Science : 109(2), 488 (1986)

6. Jenness P. K. : " Rheological design using Laponite and Cellulose Ethers", available from Laporte Absorbents Europe.

The Effect of Driers in Water-borne, Oxidatively Drying Surface Coatings

J. H. Bieleman

SERVO DELDEN BV, PO BOX 1, NL-7490AA DELDEN, THE NETHERLANDS

SUMMARY

Driers are added to unsaturated binder systems to speed up the drying process by catalysing autoxidation reactions during film formation. The composition of water-borne surface coatings differs greatly from the traditional, white spirit-thinnable alkyd paints, especially in respect of solvents and binders and the use of neutralising agents. Furthermore, since the drying of water-borne coatings is accompanied by transition from a polar to a non-polar phase, it is necessary to adjust the drying system, in respect of composition and metal concentration, to these special requirements. Pre-complexed, water emulsifiable driers are characterised by their excellent optical properties and their effective stabilisation of the drying action.

The most effective driers are the primary driers, such as cobalt and manganese. They override secondary driers more strongly in aqueous solution than in solvent-borne alkyd paints. The influence of secondary driers on the rate of drying is strongly system-dependent.

1.   INTRODUCTION

The binder in water-thinnable, oxidatively drying surface coatings consists in most cases of an alkyd resin emulsion or of a mostly colloidally dispersed alkyd resin, combined with a physically drying polymer dispersion. Drying takes place physically by evaporation of the water and of the solvents and neutralising agents which are still present, and by subsequent oxidative polymerisation of the alkyd resin. This polymerisation process is identical to that taking place in solvent borne systems and is appreciably speeded up by driers (1, 2, 3).

There are, however, distinct differences, due to the characteristic properties of water and the neutralising agents (4, 5). Moreover, in the water borne systems a transition takes place from the aqueous to the organic phase in the course of the drying process. This has a great influence on the drier and explains the differences in the drying behaviour of water borne and solvent borne coatings (6). Since the very first attempts to prepare water-borne alkyd paints, it has been known that the addition of driers can lead to problems, such as indicated in table 1.

Table 1.   Problems with Air-Drying WB-Coatings

- Poor initial dry
- Drier inhibition on storage
- Drier incompatibility with resin
- Surface defects
- Loss of stability of the colloidal system
- Loss of gloss

As a result of developments in water-borne alkyd resin systems and pre-complexed water emulsifiable driers, it has now become possible to produce water-borne paints having properties comparable with those of solvent borne paints.

This article discusses driers for water-borne surface coatings and any possible reciprocal effects between the driers and other additives.

2.   DRIERS

DIN 55 901 defines driers as follows: "Driers, also referred to as siccatives when in solution, are organometallic compounds soluble in organic solvents and in binders. Chemically they belong to the class of soaps and they are added to unsaturated oils and binders in order to appreciably reduce their drying times, i.e. the transition of their films to the solid phase. Driers are available either as solids or in solution. Suitable solvents are organic solvents and binders. Water-emulsifiable driers may contain emulsifiers".

Stewart (10) tested 35 different metal soaps as driers. Only 10 of the tested compounds showed a more or less accelerating effect on the drying process. The most favourable properties were shown by cobalt and, to a distinctly lesser degree, manganese.

Driers are usually subdivided into the following groups (8,9):

- *Primary driers*: soaps of metals which exist in several oxidation states and therefore undergo a reduction reaction. These include cobalt (Co), manganese (Mn), vanadium (V) and cerium (Ce).

- *Secondary driers*: soaps of metals which exist in a single oxidation state only and which have a catalytic effect only in conjunction with primary driers. These include calcium (Ca), zinc (Zn) and barium (Ba) and strontium (Sr).

- *Co-ordination driers*: driers are called co-ordination driers, e.g. zirconium, if the action of the metals in enhancing the drying process is based on a reaction with hydroxyl or carboxyl groups in the binder.

3.   AUTOXIDATION AND FILM FORMATION

Film formation of oxidatively drying binding media is based on molecular enlargement and crosslinking, initiated by oxidative processes. The resultant macro-molecules have a lattice structure and form insoluble and infusible films. To speed up this reaction catalytically, driers are used.

The induction period (table 2) occurs because the effects of solvent and natural anti-oxidants that may be present in the resin must be overcome before the drying process can begin.

Oxygen is then adsorbed from the air at the unsaturated sites of the resin molecule, and the adsorption continues. This process is catalysed by drier metals, dominated by cobalt.

Autoxidation of the isolenic acids in the alkyd binder first gives rise to hydroperoxides with uptake of atmospheric oxygen. As a result of the catalytic effect of the metal ions in the drier, peroxyl radicals form from the hydroperoxides, (table 3). These radicals initiate the polymerisation of the unsaturated molecules of the binding medium (3, 8, 13). Oxygen-carbon and carbon-carbon bonds are formed. Polymerisation causes gelling of the film, followed by drying and hardening. The film hardness is determined by the number of crosslinking sites. The rate of drying is strongly dependent on the action of the drier.

Table 2.   **Drying by Autoxidation**

1.   Induction period
2.   Oxygen absorption
3.   Peroxide formulation
4.   Peroxide decomposition to free radicals
5.   Cross-linking

Table 3.   **Cobalt Catalysed Drying**

$$R - O - O - H + Co^{2+} \longrightarrow RO^\bullet + Co^{3+} + OH^-$$
$$R - O - O - H + Co^{3+} \longrightarrow ROO^\bullet + Co^{2+} + H^+$$

4.   DRIERS FOR WATER-BORNE COATINGS

4.1.   COORDINATION EFFECTS ON DRIERS

Essentially the same mechanism for drying applies to both water-borne and solvent-borne air-drying vehicles. Nevertheless the drying performance is quite different. In addition to the solvent composition, the vehicle system is responsible for various drying deficiencies with water-borne coatings, such as slow dry time, loss-of-dry, poor through-drying, and hardness.

Water may hydrolyse the vehicle resulting in loss-of-dry. Water can also slow the oxygen uptake of the vehicle thereby slowing the auto-oxidation process. Water also effects the stability of the drier. Being a strong ligand, it complexes metal ions such as a cobalt. The resulting $Co(OH_2)^{2-3}$ - complex has a weaker oxidation potential, resulting in a reduction of the performance of cobalt as an autoxidation catalyst, furthermore, it is unstable. A practical remedy to compensate the loss-of-drier catalyst through hydratation is to use very high levels of the primary drier - cobalt or manganese. Compared to traditional solvent-based coatings, the level of the primary drier may be doubled, so up to 0.1 - 0.15% Co/Mn drier metal on resin solids.

A further improvement is obtained using pre-complexed driers, ligands such as o-phenanthroline or 2,2 bipyridyl (11, 12) have been known as drier activators even for traditional solvent-based paints. However, their high cost and specific application field have limited their use largely to solvent-borne polyurethane application.

Also in water-borne coatings, ligands have been used to increase drier activity and to avoid loss-of-dry. Ligands effect the spin multiplicity of electron distribution in the metal. Both stability (thermodynamic) and rate (kinetic) factors are influenced by the ligand. Water is a typical weak field ligand which tends to yield high spin complexes (figure 1).

The ligand structure can have a very important influence on the rate of electron transfer.

Fig. 1.   d Electron Distribution

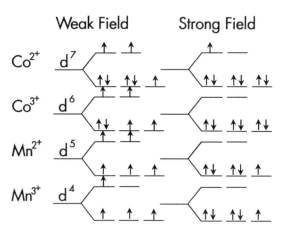

Electron distribution

For example $(Co^{111}(phen)_3)^{3+}$ is reduced more rapidly than
$(Co^{111}(NH_3)_6)^{3+}$ or $(Co^{111}(ethylene\ amine)_3)^{3+}$.

The relative low energy vacant anti-bonding orbitals found in the conjugated phen facilitate electron transfer.

Thus the effectiveness of the drier catalyst can be increased by choosing the right ligand for the drier metal ion. The fact that both oxidation states of the primary driers cobalt and manganese must be present to function effectively makes it very complicated to determine which ligand properties should be optimized for drying optimization. Adding solubility constraints complicates the issue further. The role that these ligands play is mainly to keep the active drier in its more active valency state, although its influence on the kinetic factors is not fully understood.

Auxiliary driers too, offer a second metal that is capable of maintaining multiple oxidation states, and can donate ligands to the primary drier metal and keep this in its more active valency state, thus increasing the activity of the primary drier.

4.2. WEB DRIERS

Conventional driers in water-borne coatings result in poor performance properties, as discussed earlier. Pre-complexation of the primary driers with ligands optimise its performance. Using ligands that have emulsifying properties for the primary drier in water results in a second advantage: apart from the improvement in the catalytic effectiveness, also the compatibility of the drier with the water-borne coating is remarkably improved.

The general structure for this new class of drier is given in fig.2.

Fig. 2.

WEB DRIERS

$(Me(Lig.)_n)^m \quad (OOCR)_m$
Me = Metal
Lig. = Ligand
n   = Number Ligands
m   = Valency Metal-io

5. EXPERIMENTAL SECTION

Drying tests were performed in order to investigate the influence of the drier in relation to the composition of the paint. Different water-borne binders were used for these paints. (See table 4.)

The first 3 alkyd binders (A, B and C in table 5) have an oil length of approx. 40-45. Also both urethane alkyd emulsions (D and E in table 6) have a similar oil length, according to the information from the supplier.

Table 4.   **Coating Formulations: binder compositions**

A.   Alkyd Emulsion/Polyacrylic Dispersion
B.   Colloidal Alkyd/Polyacrylic Dispersion
C.   Alkyd Solution
D.   Medium Oil Urethane Alkyd
     Emulsion - Supplier A
E.   Medium Oil Urethane Alkyd
     Emulsion - Supplier B

Table 5.   Composition of the Surface Coatings

Composition	A (%)	Paint B (%)	C (%)
HEC (2% solution in water)	24.0	13.1	-
Dispersant	0.2	0.7	-
Wetting agent	0.25	0.3	-
Antifoam	0.25	0.1	-
Water	1.8	15.8	-
Alkyd resin (75% solution in butyl glycol and n-butanol 1:1)	-	-	27.2
Titanium dioxide	18.0	23.0	27.2
Disperse and add in the order stated:			
Alkyd resin (50% emulsion in water)	31.0	-	-
Alkyd resin (75% emulsion in water)	-	28.1	-
Acrylic resin (46% dispersion in water)	23.0	-	-
Acrylic resin (50% dispersion in water)	-	14.0	-
Antifoam	-	0.1	-
Corrosion inhibitor	-	0.2	-
Biocide	-	0.1	-
EGMBE	-	-	2.7
n-Butanol	-	-	3.1
Ammonia to pH 8.2-8.5	-	-	1.0
Water	-	-	38.8-a
Drier	1.5	0.9	a
Water	-	3.6	-
Total	100	100	100
Solid binder	26	28	27
PVC	14	18	20

Table 6:   Composition of the WB primers

	D	E
Water	18	
Dispersant	1	
Anti-foam	0.7	
Butanol	1.5	
DMEA	0.5	
Iron Oxide	5	11
Zinc Phosphate	10	11
Talcum	5	6
Extender	5	11
Alkyd Emulsion (50% solids)	8	25
Disperse and add:		
Alkyd Emulsion (50% solids) 26	30	
Defoamer	0.3	
PUR-thickener (SER-AD FX 1010, 3%)	10	
Water	9-d	6-e
Drier	d	e
Total	100	100
Solid binder content of primer	17%	27.5%

Table 7:   Driers

Type	Metal content (%)	Solubility	
		White Spirit	**Water**
Co octanoate 10%	10	readily soluble	insoluble
WEB Co 8%	8	readily soluble	emulsified
Co-Ba-Zn	11.6	readily soluble	insoluble
WEB Co-Ba-Zn	11.6	readily soluble	emulsified
WEB Mn 9%	9	readily soluble	emulsified
WEB Co Special	17	dispersed	dispersed

5.1   DRIERS USED

Three different classes of driers were chosen (table 7).

A.   Octoates, i.e. traditional, non-water soluble or self-emulsifiable driers.

B.   A pre-complexed metal drier, soluble in white spirit but not soluble or self-emulsifiable in water. This drier coded as FS 530 is composed of manganese complexed with a strong field ligand.

C.   WEB driers, a unique class of pre-complexed metal driers, emulsifiable in water and soluble in white spirit. Their solubility in hydrocarbon solvents (and polymers) is due to the hydrocarbon chain in the anionic part of drier. Their emulsifiability in water is due to the polarity of the ligand and/or the presence of emulsifiers. (figure 2)

The complete study included both single driers as well as combination driers within these 3 classes. The most relevant results are reported here.

PROCEDURE

The drying time of the paint was determined using a Beck-Koller drying time recorder. Pendulum hardness was measured by the König method and gloss by the Erichsen method. All tests were performed in a constant temperature and humidity chamber at 20°C and 65% relative atmospheric humidity. Other optical properties were assessed visually. Each measurement was carried out both 24 hours after preparation and after storage for 6 months at 20°C.

6   RESULTS

The results are summarised in tables 8 to 12. It follows from the test results in the top coatings that best optical properties and gloss is produced using the water-emulsifiable pre-complexed driers. The octoates are responsible for coagulation or flocculation phenomena (see tables 8, 9), observed in the wet paint layer just after application. This also leads to an irregular coating appearance of the dried layer. In the top coating, prepared with the colloidal alkyd solution (formulation C) as well as in both primers (formulations D and E) no differences were detected between the water-emulsifiable and the non-emulsifiable driers.

The effect of the class of driers on the drying properties is apparently quite dependent on the composition of the coating. In the top coating according to formulation A we noticed a strong improvement of the initial drying time used the WEB drier. However in formulation C no improvement could be found.

Similar effects of the class of drier as on the drying properties, have been obtained for the influence on the hardness.

Obviously the drier metal concentration on resin solids is of great importance for the drying time. Doubling the amount of cobalt from the traditionally used level of 0.05 to 0.1% results in formulation B to an almost twice faster drying time. However, hardly no influence of doubling the primary drier concentration is seen in the two primer formulations.

Also, regarding the influence of auxiliary driers, no uniform effect could be determined. In some formulations a strong, positive effect on the rate of drying is noticed (e.g. in test formulation A), in others (test formulations D and E) no effect has been registered.

The effectiveness of the pre-complexed manganese drier FS 530 is very selective: is most effective in the primer formulation B, but not in the primer formulation A, both based on apparently similar classes of binders.

For all cases no severe drying problems, i.e. loss-of-dry, is noticed after storage of the paint during several months.

Table 8:   **Test results on Formulation A**

Drier	Metal (%) to solid binder	Drying time (h) (colour coat)		König hardness (s)	
		Immed.	after 6 mths	Immed.	after 6 mths
Co octoate 10%	0.1 Co	8.30	>8	31	29
WEB Co 8%	0.1 Co	5.30	>8	33	29
Co-Ba-Zn octoate	0.1 Co 0.6 Ba 0.27 Zn	2.00	8	29	29
WEB Co-Ba-Zn	0.1 Co 0.6 Ba 0.27 Zn	2.00	8	31	29

Drier	Gloss 60°		Appearance		pH	
	Immed.	after 6 mths	Immed.	after 6 mths	Immed.	after 6 mths
Co octoate 10%	48	22	coag.	coag.	8.5	7.7
WEB Co 8%	48	45	good	good	8.4	7.8
Co-Ba-Zn octoate	24	23	coag.	coag.	8.4	7.8
WEB Co-Ba-Zn	27	24	good	f-good	8.5	7.8

Table 9:   **Test results in Formulation B**

Drier	Metal (%) to solid binder	Drying time (h) (colour coat)		König hardness (s)	
		Immed.	after 6 mths	Immed.	after 6 mths
WEB Co 8%	0.05 Co	>7	>7	17	20
WEB Co 8%	0.1 Co	4	7	20	22
WEB Co-Mn	0.05 Co 0.1 Mn	>7	>7	21	21
Co-Ba-Zn octoate	0.1 Co 0.6 Ba 0.2 Zn	5	5	14	18
WEB Co Special	0.1 Co	4.5	4	31	30

Drier	Gloss 60°		Appearance		pH	
	Immed.	after 6 mths	Immed.	after 6 mths	Immed.	after 6 mths
WEB Co 8%	53	44	good	good	6.5	5.4
WEB Co 8%	57	42	good	good	6.0	5.3
WEB Co-Mn	52	41	good	good	6.3	4.7
Co-Ba-Zn octoate	33	22	floc.	floc.	5.6	4.7
WEB Co Special	47	43	good	good	6.6	5.0

Table 10:   **Test results in Formulation C**

Drier	Metal (%) to solid binder	Drying time (h) (colour coat)		König hardness (s)	
		Immed.	after 6 mths	Immed.	after 6 mths
WEB Co 8%	0.05 Co	0.5	3.5	67	35
WEB Co-Mn	0.05 Co 0.1 Mn	0.4	3.5	68	57
WEB Co Special	0.1 Co	0.5	2.0	67	57
Co-Mn octoate	0.05 Co 0.1 Mn	0.4	3.5	68	49

Drier	Gloss 60°		Appearance		pH	
	Immed.	after 6 mths	Immed.	after 6 mths	Immed.	after 6 mths
WEB Co 8%	87	81	good	good	8.5	7.8
WEB Co-Mn	91	85	good	good	8.6	7.7
WEB Co Special	90	81	good	good	8.6	8.2
Co-Mn octoate	90	80	good	good	8.6	7.7

Table 11:   **Test results in Formulation D**

Drier	Metal (%) to solid binder	Through-drying h/120 μm wet		König hardness (s)		Appearance	
		Immed.	after 2 mths	Immed.	after 2 mths	Immed.	after 2 mths
FS 530*	0.1	6	12	9	8	O.K.	O.K.
WEB Co 8%	0.05	4	6	29	32	O.K.	O.K.
WEB Co 8%	0.1	4	5	32	36	O.K.	O.K.
WEB Co/Zr	0.05/0.1	4	6	30	32	O.K.	O.K.
None	-	>24	>24	7	7	O.K.	O.K.

* Manganese complexed drier

Table 12:   Test results in Formulation E

Drier	Metal (%) to solid binder	Through-drying h/120 µm wet		König hardness (s)		Appearance	
		Immed.	after 2 mths	Immed.	after 2 mths	Immed.	after 2 mths
FS 530*	0.1	4	5	17	18	O.K.	O.K.
WEB Co 8%	0.05	7	20	18	13	O.K.	O.K.
WEB Co 8%	0.1	4	15	19	16	O.K.	O.K.
WEB Co/Zr	0.1/0.1	4	15	19	16	O.K.	O.K.
None		>24	>24	6	6	O.K.	O.K.

* Manganese complexed drier

7.   DISCUSSION

7.1.   OPTICAL PROPERTIES AND DRYING CHARACTERISTICS

No uniformity in the influence of one class of drier on the drying properties of the various coating formulations could be determined. Obviously there is a strong dependency of the effectiveness of the drier on the composition of the coating formulation. No single class of drier will give the most optimal drying properties in all coatings. However, this does not count for those properties that are not directly related to the catalytic effect of the drier. Optical properties, paint appearance, as well as gloss are in all cases best or equal to the best when water emulsifiable WEB driers are being used. Apparently this class of driers shows best compatibility with both phases involved and do not cause pigment-, binder-flocculation or coagulation. The difference between the different classes of the driers in influence on the optical coating layer properties is not of the same level for all coating formulations. The greatest differences are noticed in formulations A and B. Both formulations show relatively weak emulsifying properties, through the water- soluble binder. The octoate drier is easily distributed in the aqueous phase by this binder. In the low-gloss primer surfaces no differences in appearance could be noticed.

Regarding the influence of the drier catalyst on the drying properties and the development of the hardness of the coating layer, it is quite evident that the primary driers are strongly dominating. A positive influence of auxiliary driers is only determined in formulation A. Besides, auxiliary driers tend to cause instability of the coating formulation, e.g. in systems containing anionically stabilised polymeric dispersions. The auxiliary drier cations such a barium, lead or zirconium may neutralise the anion and thus loose its properties as a drier as well as destroying the stability. So in such systems, auxiliary driers should be chosen with care and certainly first be evaluated on compatibility and stability. The level of the primary driers, cobalt and manganese that give the best drying properties is much higher than is normally used in traditional solvent-based coatings. This is also the case when using pre-complexed driers.

A reason for this can be the lower mobility of the drier in the aqueous phase. Different from solvent-based systems, the drier is not soluble in the continuous phase but emulsified directly or indirectly (as solution in the resin emulsion particles) in these. So it may be assumed that during the drying period less contact areas may exist between the drier and the resin molecules. This is to be outweighed with a higher concentration of the drier. Part of the drier may also be lost due to various reactions and interactions between the drier and other components in the paint system (see 7.2). No clear explanation could be found for the differences in performance of the driers in the primer formulations D and E.

Apparently these systems contain similar binders. It may be assumed that the stability of the manganese complex in FS 530 is effected by the DMEA, used as neutraliser in formulation D. This amine may replace the ligand in FS 530, resulting in lower catalytic effectiveness.

7.2 RECIPROCAL EFFECTS BETWEEN DRIERS AND OTHER PAINT COMPONENTS

7.2.1 WATER

From an ecological standpoint, water is a very attractive solvent and thinner. However, when combined with driers, it has some less attractive physical characteristics. Due to its polarity, water is an excellent solvent and a good reaction medium for ionised and polar compounds, such as the metal ions of the driers and the carboxyl ions of the binding medium. (table 13)

Table 13:   Water

- Polar Solvent
- Reaction Medium
- Ligand
- Low Oxygen Solubility
- High Latent Heat of Evaporation
- High Surface Tension

Moreover, water is one of the strongest complexing agents with metal ions, such as Co^{3+} and Mn^{3+}.

Finally, mention must be made of the tendency to hydrolysis. The hydrate $(Co(H_2O)_6)^{3+}$ is unstable and is almost spontaneously hydrolysed.

Table 14:   Influence of Emulsifiers on Drying and Hardness

Paint system: medium-oil alkyd resin, 60% solutions in white spirit. Amount of emulsifier: 5% solid on solid.

Emulsifier	Drying time (h)	Koenig (s)hardness
none	6.30	55
NP-14 (1)	17.00	35
AAS (2)	7.30	53
Ca DBS (3)	24.00	54

(1)   nonylphenol polyethylene glycol, 15 ethylene oxide units
(2)   ammonium alkylsulphate
(3)   calcium dodecylbenzensulphonate

7.2.2 BINDERS

Binders for air-drying water-borne paints include:

- alkyd resin emulsions
- self-emulsifying alkyd resins
- water-soluble alkyd resins

The emulsions mostly contain nonionic emulsifiers and are stabilised by protective colloids. Although these resins differ only to a slight extent from solvent-borne alkyd resins, they have the drawback of impairing the properties of the film by their high emulsifier content. Table 14 demonstrates the influence of emulsifiers on the rate of drying and film hardness of alkyd resins in white spirit solution in the presence, in each case, of 5% of emulsifier and of driers comprising of 0.05% Co - 0.3% Zr - 0.1% Ca. The results indicate that the drying rate and hardness are impaired by nonionic emulsifiers (nonylphenol polyethylene glycol - 15 ethylene oxide units), but are hardly affected by ammonium alkyl-sulphate.

Self-emulsifying alkyd resins contain hydrophilic groups and can therefore be emulsified in water without the addition of emulsifiers. Polyethylene glycol units are incorporated into the molecule as hydrophilic groups, in addition to acid groups. The polyethylene glycol groups thus have an adverse effect on the rate of drying and film hardness, just as the nonionic emulsifiers have in the case of emulsions.

Water-soluble alkyd resins are true or colloidal solutions of alkyd resins with a high acid value which have been neutralised with ammonia or amines (7, 23). The acid value increases on storage owing to hydrolysis. The solubility and effectiveness of driers may be influenced by salt formation (14).

Table 15: **Influence of the Neutralising Agent on the Drying Time of a Water-borne Alkyd Paint**

% Metal on binder solid	Drying time(h) using:		
	none	ammonia	triethylamine
0.2% Co	3.5	12	10
0.2% Mn	6	8	2

7.2.3 NEUTRALISING AGENTS

Nitrogenous neutralising agents, such as ammonia or amines, form with cobalt or manganese complex metal ions which have a slight catalytic action. Cobalt is influenced to a greater degree by ammonia and manganese by amines (Table 15).

7.2.4 PIGMENTS

Driers have a certain boundary surface affinity with pigment substrates. Driers are adsorbed by the pigments, as they are in solvent-borne paints (16). This adsorption, however, has no influence on the action of driers (17).

8.   APPLICATION AND DRYING CONDITIONS

Next to the composition of the coating are also the application and drying conditions of paramount import-ance for an optimal result. During the application of water-borne paints it is important that the relative humidity is not too low and not too high. A too low humidity will result in e.g. poor levelling, short overlap time, brush drag or dry-spraying, (when using airless spray equipment). The lower the relative humidity, the faster the water will evaporate. For optimal application and drying, a relative humidity of preferably between 50 and 80% is needed.

9.   CONCLUSION

The differences in formulation and in the film-forming process between aqueous systems and conventional alkyd paints make accurate adjustment of the drier system essential. The use of water-emulsifiable, pre-complexed WEB driers improves the optical properties of the films, especially the gloss, and reduces surface defects. Depending on the formulation, the rate of drying is also improved, especially after storage. The most effective driers are the primary driers, such as cobalt and manganese. They override secondary driers more strongly in aqueous solution than in solvent-borne alkyd paints. The influence of secondary driers on the rate of drying is strongly system-dependent.

References

(1)	H. Schumacher, Double Liaison (1986), 399
(2)	J.W. Nicholson, JOCCA 1 (1987), 1
(3)	J.H. Bieleman, Pol.Paint Colour J., vol.182 (1992), 412
(4)	A. Jones, L.Campey, J.Coat.Tech., vol.56, 713 (1984), 69
(5)	R. Zimmermann, Farbe & Lack, 89 (1983) 7, 499
(6)	J.W. Nicholson, Occa Monograph no. 2 (1985), p.1
(7)	E. Schulze, Polym.Paint Col. J., vol.175, 4145 (1985),415
(8)	Ullmanns Encyklopdie der technischen Chemie, 4,vol.23, 421
(9)	Solmon, The Chem.of organic Film Formers, John Wiley, New York (1981), p. 398
(10)	W.J. Stewart, Off. Digest 26 (1954), 413
(11)	Middlemiss, J. of Water Borne Coatings (1985), 3
(12)	Belletiere, Mahoney Journ.Coat.Techn. 59 (1987), 752, 101
(13)	J. Halpern, Collect.Pap.Symp.Coord.Chem. (1964), 351
(14)	J. Engel, J. Water Borne Coat. Aug. (1983), 3
(15)	R. Hurley. F. Buono, J. Coat. Techn., vol.54, no.694 (1982), 55
(16)	J.H. Bieleman, Polym.Paint Colour J., 6 (1982), 423
(17)	S. Wislka, O. Salminen, Paint & Resin, no. 9 (1982), p.8

A Silver Lining for Paints and Coatings – A Revolutionary Preservative System

K. D. Brunt

MICROBIAL SYSTEMS INTERNATIONAL LIMITED, GOTHIC HOUSE, BARKER GATE, NOTTINGHAM NG1 1JU, UK

The antimicrobial activity of silver has been known for a very long time – certainly from the days of the ancient Romans, but until very recently it has not been possible to make use of this property for general industrial applications. In ancient times the wealthy kept drinking water in silver pitchers, as it enabled those so minded to drink water rather than alcohol–based refreshments but with a reduced risk of intestinal eruptions. More recently, a solution of silver nitrate has been dropped into the eyes of new–born infants to prevent post–natal bacterial infection of the eyes. Over the last ten years or so, the use of silver coated onto sand or carbon has become common for use in potable water treatment. The problem is that although silver's qualities of antimicrobial potency and human safety have become obvious through these applications, it has not until now been possible to harness them for industrial preservation.

For preserving liquid products, silver filings, be they ever–so finely divided, would be clearly no good: they would settle out very quickly, be relatively inactive but cost a great deal. Silver nitrate would also be a disappointment – it would exhibit a brief antimicrobial activity and then proceed to turn the preserved products black! The trick, it has been discovered, for liberating silver's antimicrobial properties, lies in the controlled release of silver ions.

This "trick" has been discovered through research done by Johnson Matthey PLC, the senior partners in the two–company co–operation developing the technology and new product range. Johnson Matthey are world leaders in precious metals technology, and have for many years been involved in the bio–medical applications of precious metals, developing such valuable pharmaceutical actives as Cis–Platin. In response to requests from the medical industry, they researched into the use of precious metals, and especially silver, in the prevention of microbial settlement on prosthetic implants, and from this programme emerged a system now in clinical trials, for coating catheters and canulae with latex bearing a composite of silver chloride on titanium dioxide. This system prevents the harmful ingress to the patient's body of pathogenic microbes along the surfaces of such implants.

This system employs the principal of controlled silver ion release in such a way as to constantly maintain a microbicidal concentration of silver ions in the aqueous environment of the composite. The composite consists of insoluble silver chloride bonded onto the surface of specially–synthesised, high porous particles of titanium dioxide (see Figure 1).

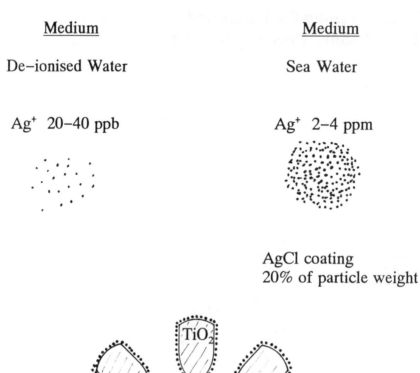

Medium   Medium

De–ionised Water   Sea Water

Ag⁺ 20–40 ppb   Ag⁺ 2–4 ppm

AgCl coating
20% of particle weight

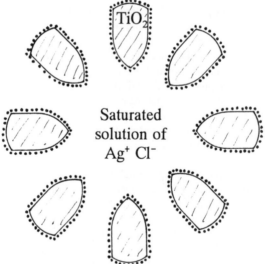

TiO₂

Saturated
solution of
Ag⁺ Cl⁻

SILVER CHLORIDE/TITANIUM DIOXIDE
COMPOSITE PARTICLE

Figure 1, The controlled release of silver ions

When this composite, of which the particles are 3 to 5 microns in diameter and contain 20 to 30% by weight of silver chloride, is put into an aqueous environment, a small amount of silver ions are released into the water. The concentration released will stabilise at an equilibrium level of 20–40ppm in de–ionised water, up to 2–4ppm in sea-water, with the rest of the silver still locked up safely in the body of the composite particles. Only when free silver ions are removed from the equilibrium by chemical combination with microbes or other active entities, will more silver ions be released from the composite – and then only enough to regain the appropriate equilibrium concentration.

This basic composite exhibits a good spread of antimicrobial activity, as demonstrated by Table 2 which gives the minimum inhibitory concentrations of the composite for various organisms.

Table 2, Minimum Inhibitory Concentrations by Agar Dilution Method

ORGANISM	MIC (mg/ml JMAC)
Staph aureus (NCTC 10788)	160
Lactobacillus buchneri	160
Pseud aeruginosa (NCTC 6749)	160
E coli (ATCC 8739)	160
C albicans (NCYC 597)	160
C albicans (NCPF 3179)	160
Sacc cerevisiae (NCYC 200)	160
A niger (IMI 17454)	160
A fumigatus (IMI 134735)	>160

Bacterial medium was Iso–Sensitest agar
Fungal medium was PA + YNB

When this data became available, it became clear that the system had further potential as an industrial preservative, and it was at this point that the second collaborator in the project becomes involved. Microbial Systems International Limited is a small company which specialises in the development and marketing of industrial antimicrobial systems, and to them fell the task of turning the composite into a commercial product range. Their initial work in testing the composite in cosmetic formulations (see Table 3) clearly show that the system had potential as a preservative in such high–value products, but that for general industrial applications it had a number of drawbacks.

The drawbacks uncovered in these initial trials were (see Table 4) in the areas of spectrum of activity, handlability, and cost.

Microbial Systems were able to apply their biocide enhancement technology so as to overcome all these problems. By suspending the insoluble composite in a water/surfactant gel, it was possible to formulate a product which was much easier to handle than an expensive microfine powder. By using a carefully selected

Table 3, Initial Challenge Test Results (BP 1988 Test Method)

TEST FORMULATION	JMAC LEVEL	RESULT
Pharmaceutical Cream Base	1.0%	Bacterial challenge eradicated by 14 days Fungal challenge not eradicated
Pharmaceutical Cream Base	0.1%	Bacterial challenge eradicated Fungal challenge not eradicated
Shampoo Base	1.0%	All challenge inocula eradicated by 48 hours
Shampoo Base	0.1%	All challenge inocula eradicated by 48 hours

Table 4, Drawbacks of the Silver Composite as a Practical Biocide

1.	Expensive
2.	Incomplete Spectrum
3.	Very Fine Powder

sulphosuccinate salt as the gelling agent, it was possible to improve the way in which the silver ions were able to penetrate into microbial cells, and thus complete the spectrum of microbiological activity and significantly enhance the cost effectiveness of the system (see Table 5).

This formulated system has now been extensively tested in a wide range of synthetic and natural latex emulsions (see Table 6) using a variation of the British Pharmacopoeia Challenge Test (1988), and has shown excellent preservative activity.

The activity of the system is so good that when trial marketing was commenced some months ago it was discovered that the system was too potent for the potential customers to handle easily, so the prototype formulation designed for use at 100ppm to 200ppm had to be reformulated so that it could be used at 1000ppm to 2000ppm, levels which many factories find easier to measure!

Apart from its excellent activity over the entire microbiological spectrum, the outstanding feature of the silver composite system is its incomparable safety in use. It is probably true to say that, apart from the view–point of
microbes, this is the world's safest effective industrial preservative. To humans, the ingredients of the preservative formulation are almost inert to the user:

1. Silver chloride – insoluble in water, will pass unchanged through the gut.
2. Titanium dioxide – as above, and widely used in cosmetics as well as paints.
3. Sodium dioctyl sulphosuccinate – used in pharmaceuticals, cosmetics and toiletries, latex emulsions and many other industrial applications. Major human hazard is that if ingested it could cause severe foaming on passing wind.
4. Water – relatively inert to humans unless in excessive quantities, widely used industrial solvent.

Table 6, Enhanced Silver Composite – Activity by BP (1988) Challenge Test Method

FORMULATION TYPE	ENHANCED JMAC CONCENTRATION	RESULT
Latex Concrete Additive	60 ppm	Poor control of bacteria
Latex Concrete Additive	150 ppm	Slow but effective control
Latex Concrete Additive	380 ppm	Effective control
Paper–coating Latex (MSI 864)	200 ppm	Effective control
Paper–coating Latex (MSI 867)	200 ppm	Effective control
Paper–coating Latex (MSI 870)	200 ppm	Effective control
Paper–coating Latex (MSI 873)	200 ppm	Effective control
Paper–coating Latex (MSI 876)	200 ppm	Effective control
Textile Additive Latex (MSI 755)	50 ppm	Slow control of fungi
Textile Additive Latex (MSI 757)	175 ppm	Satisfactory control of all groups

Table 5, Enhanced Silver Composite System MIC's by Agar Dilution Method

ORGANISM	MIC (mg/ml Enhanced JMAC)
S epidermis biotype 3	100
Ps aeruginosa (NCTC 6749)	100
E coli (ATCC 8739)	200
B subtilis (NCTC 3160)	36
B cereus (NCTC 7464)	52
C albicans (NCPF 3179)	52
Sacc cerevisiae (NCYC 200)	200
Oidium sp	52
A fumigatus (IMI 134735)	13
Tricoderma viride	13
Acternaria alternata	13

A full toxicological programme is under way on this novel system, and the key results are summarised in Table 7. One of the greatest difficulties in the testing programme has been that of finding toxic effect levels at all, as distinct to the usual nervous establishment of "no-effect" levels.

Some worries have been expressed from the environmental point of view, regarding the possibility of view the silver system as a "heavy metal". An examination of the levels of silver involved, however, shows that they are so low as to pose no environmental hazard. This may be demonstrated by the simplified spill-dilution example shown in Table 8.

Should quantities of the biocide itself be spilt, it can be cheaply and effectively "killed" by the application of sodium metabisulphite, which will reduce the active silver to inert silver sulphide.

A programme is in place to secure a full set of legislative approvals. As may be seen from Table 9, approval for use in cosmetics and toiletries has already been granted in the USA by the CTFA, and EC cosmetic approval (a rather more bureaucratic procedure) is expected for the second half of 1995. Approval for use in food contact

Table 9, Silver Composite System – Approvals

BGA	Food contact approval forecast for November 1994
FDA	First food contact category forecast for November 1994
CTFA	Approved for skin cosmetics and rinse-off products
EC	British Prior National Approval forecast for January 1995 Full EC Cosmetic Ingredient Listing forecast for late 1995

Table 7, Safety of the Silver Composite System

STUDY	DOSE LEVEL	RESPONSE
Acute Oral	5000 mg/kg	No effect
Intraperitoneal	50 mg/kg in saline	No effect
Intravenous	8.6 mg/kg	No effect
Inhalation	200 mg/litre	No effect
Eye irritation	1% in cream	No effect
Eye irritation	100% powder	Mild/moderate irritant
Skin irritation	100% powder	No effect
Human irritation/ sensitisation	1% in cream	No effect
Photosensitisation	1% in cream	No effect
Sensitisation (Magnusson & Kligman)	25% in water 50% in water	No effect No effect
Migration – Human skin	1%	No recorded migration through dermis at limit of 3ppb
Migration – latex coating on paper	100 ppm	Below limit of detection of 3 ppm aqueous and fatty media

Table 8, Silver Composite – Environmental Impact

Silver chloride/titanium dioxide composite	Approximately 20% AgCl w/w
Enhanced silver chloride	Approximately 10% silver composite w/w i.e. about 1% silver w/w
Emulsion paint containing 0.1% enhanced composite	About 0.01% silver
S P I L L !	
Hose away to drain	Say 1000:1 dilution – 0.00001% silver
Further dilution in sewer	Say 1000:1 dilution – 0.00000001% silver i.e. 100 ppb
Typical silver content of garden soil (UK)	20 – 200 ppb

applications may well be granted by the German BGA and American FDA by the time this paper is in print, and the full EPA approval programme is now running and expected to take about two years.

In summary, the method has been discovered to unlock the remarkable microbicidal potential of the silver ion, and this has been developed into a family of formulations to give powerful, versatile and exceptionally safe preservation in a wide range of industrial products and processes.

Table 10, The Enhanced Silver Composite System

POWERFUL	Low use levels Broad spectrum
VERSATILE	Wide range of applications Liquid or solid systems
EASY TO USE	Readily water miscible Chemically and thermally stable
SAFE	Non–irritant Non–sensitising No hazard labels

Subject Index